ANSWER MANUAL FOR
GENETICS

ANSWER MANUAL FOR
GENETICS
Third Edition

MONROE W. STRICKBERGER
University of Missouri—St. Louis

Macmillan Publishing Company
New York
Collier Macmillan Publishers
London

Macmillan Publishing Company
866 Third Avenue, New York, New York 10022

Collier Macmillan Canada, Ltd.

ISBN 0-02-418130-7

Printing: 1 2 3 4 5 6 7 8   Year: 5 6 7 8 9 0 1 2 3 4

ISBN   0-02-418130-7

Preface

This manual was written to provide solutions to all of the problems in my book, *Genetics*, Third Edition (Macmillan, 1985), with full explanations offered wherever they might be helpful. It incorporates corrections and suggestions sent to me by numerous colleagues and students who used the first two editions. I am especially indebted to James N. Thompson, Jr., Jenna J. Hellack, and Gerald Braver, all at the University of Oklahoma, who reviewed the entire *Answer Manual* manuscript for this edition, and proposed various answers and suggestions. I will be grateful if readers will notify me of errors and ambiguities that remain, and will make every effort to have corrections incorporated into any further printings of the manual.

M. W. S.

Contents

(Chapter numbers and titles refer to the problems given in the listed textbook chapters.)

2

Cellular Division and Chromosomes

2-1. (a) Since each cell has only one pair of telocentric chromosomes, each V-shaped structure in the diagram that is being pulled to a separate pole must represent a replicated telocentric chromosome in which the centromeres of the two daughter chromatids are still connected. The diagram in Problem 2-1, therefore, represents the separation between two dyads or anaphase of the first meiotic division.

(b)

2-2. (a)

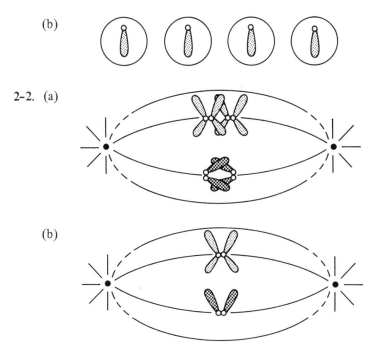

(b)

2-3. The cross-configuration is characteristic of a metacentric chromosome in metaphase, in which all chromosome arms are approximately equal in length. There are two possible explanations for the origin of this configuration:

    1. The cell is in the second meiotic division, having begun meiosis with two paired metacentric chromosomes (2n = 2), and reductional division has

already occurred, leaving the observed chromosome with its two attached chromatids to undergo a forthcoming equational division.

2. The cell represents the metaphase stage in a haploid organism or cell, which began mitosis with a single metacentric chromosome.

2-4. $A^1 A^2 B^1 B^2$

2-5. (d)

2-6. When chromosome A separates from A' at the first meiotic division, the B chromosome may go to either pole. In about half of the meioses, the four resulting gametes will be AB, AB, A', A', and in the other half they will be $A'B, A'B, A$, and A.

2-7. $ABCD$ gametes are produced. At fertilization, the zygote will have the genotype $AABBCCDD$.

2-8. (a)

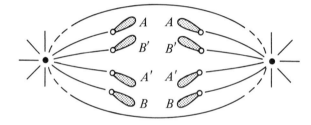

(b) The chromosomes could be arranged in either of the following two ways:

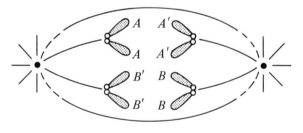

or

(c) For the first arrangement:

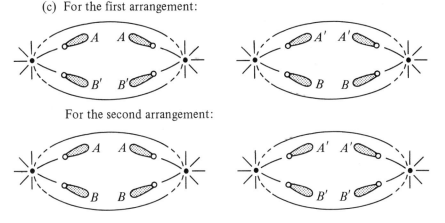

For the second arrangement:

(d) Since chromosomes divide the same way in eggs as they do in sperm, the kinds and frequencies of female gametes would be the same as the kinds and frequencies of male gametes. The *number* of female gametes, of course, would be expected to be smaller since each meiosis produces only one female gamete rather than four, and the number of meiotic divisions in females is usually less than the number of meiotic divisions in males. This reduction in numbers, however, should not affect one kind of gamete more than any other.

2-9. The probability that the first chromosome derives either from maternal *or* paternal origin is equal to 1; that is, it must come from one parent or the other. Thereafter, the likelihood that the next chromosome comes from the same parent is 1/2. Therefore, the probability that the six chromosomes in a gamete come from one parent is $1 \times 1/2 \times 1/2 \times 1/2 \times 1/2 \times 1/2 = 1/32$, which is $(1/2)^{n-1}$, where n is the number of chromosome pairs (see text p. 21). The student's statement is therefore incorrect.

2-10. Since each member of a pair of homologous chromosomes comes from a different parent, each may be carrying different genes. In meiosis, members of homologous pairs separate, thereby producing genetic differences between gametes. In mitosis, each daughter cell contains all homologues, and mitotic products therefore are expected to be genetically alike.

2-11. If we symbolize each pair by letters—for example, A (knob), A' (non-knob), and B (satellite), B' (nonsatellite)—then there are four possible kinds of gametes produced in meiosis: $AB, AB', A'B$, and $A'B'$. Or, diagrammatically,

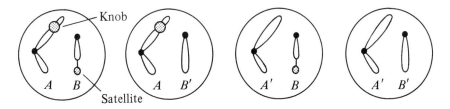

2-12. Mitosis. Since there are no homologues to pair with, sex cell chromosome division will appear similar to the second meiotic division in a diploid. (In bees, each spermatocyte divides mitotically to form two cells, one large and one small, each with apparently the same chromosome constitution. Only the large cell, however, becomes a functional sperm. This is in contrast to ants, wasps, and some other insects in which both mitotic products of the haploid male spermatocyte form functional sperm.)

2-13. Since most or all cytoplasmic particles in a zygote are derived from the egg rather than from the sperm, cytoplasmic heredity would cause greater resemblance between an offspring and its female parent.

3

Reproductive Cycles

3-1. If the first meiotic division was reductional for the centromere of this chromosome, the ascospore arrangement in the ascus would follow the pattern *AAAAaaaa*, as shown in the upper part of the accompanying diagram (p. 6). If the second meiotic division was reductional, the expected results would be *AAaaAAaa* (lower part of the diagram) or *AAaaaaAA* or *aaAAAAaa*. [It is the first meiotic division that is reductional for the centromere in *Neurospora*, whereas the second meiotic division may be reductional for genes located along the chromosome farther from the centromere (Chapter 16).]

3-2. (3)

3-3. (a) Yes.   (b) No.   (c) $AB; AB'; A'B; A'B'$

3-4. If only replicates of a single chromosome are to be separated, it should be possible to do this without necessarily involving the complexities of aster formation and the multiple protein strands used in spindles. In bacteria, for example, it is probable that separation of the daughter chromosomes occurs by their individual attachment to the cell wall and subsequent growth and movement of the cell wall so that the formerly connected points are now separated.
(See reference of Jacob, Ryter, and Cuzin, text Chapter 29, and a study by A. Worcel and E. Burgi, 1974, *J. Mol. Biol.*, 82:91-105.)

3-5. (a) 8   (b) 8   (c) 16

3-6. Organisms that reproduce *asexually* can be different from their parent and from others in the same "clone" only when a new hereditary change (mutation) occurs. Organisms that reproduce *sexually* can be different both from their parents and from others in the same family because of new combinations between already existing hereditary changes. We would therefore expect greater variability among the latter than among the former.

3-7.   (a) No difference between endosperm cells in a seed is expected because they are all replicated by mitosis, and there is no reduction division in mitosis.

5

Figure for Problem 3–1

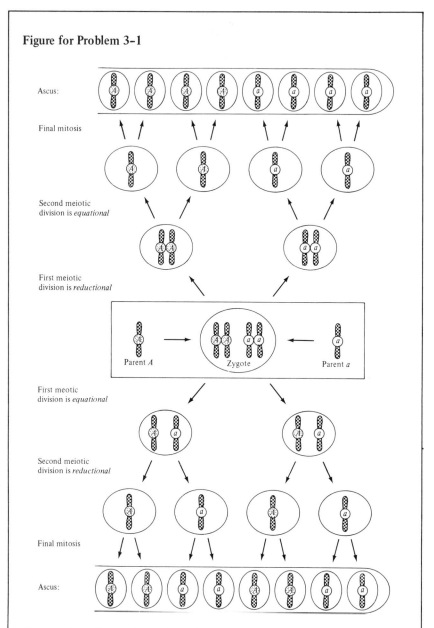

Diagram showing two hypothetical consequences for a *Neurospora* zygote (center of diagram) containing a pair of homologous chromosomes whose centromeres are labeled *A* and *a*. The upper part of the diagram shows the ascus formed by this zygote when the first meiotic division is reductional, and the lower part of the figure shows the ascus formed when the first meiotic division is equational.

(b) Differences between pollen cells formed by the triploid micro-sporocyte are expected since such cells undergo a reduction division, and either two homologous chromosomes go to one pole and the third homologue goes to the other pole, or all three homologues go to one pole and none goes to the other pole (see also text Fig. 21-1).

3-8. (a) 20   (b) 5   (c) 5   (d) 5

3-9. (a) 20   (b) 10   (c) 5   (d) 5

3-10. (a) 46   (b) 23   (c) 23   (d) 46   (e) 23

3-11. (a) 400   (b) 400   (c) $400 \times 3 = 1200$. This number however, is only theoretical because not all polar bodies are usually formed.
(d) 23

3-12. (a) $32 + 31 = 63$
(b) The hybrids are expected to be sterile since, in the absence of pairing, the reduction division in meiosis will produce gametes, each with different numbers of chromosomes (see text p. 15).

4

Nucleic Acids

4-1. (a) An organism that has a total of only four different "traits" could easily use one of the four different nucleotides (A, T, C, or G) to designate a trait. Therefore, a sequence of one nucleotide would seem sufficient to code for one of these traits (4^n, where n = 1; see text p. 44).

(b) An organism with 16 different traits obviously needs a nucleotide sequence that must have 16 possible arrangements. This can be achieved by a linear sequence of two nucleotides (4^2), for example, AA, AT, AC, AG, . . ., TA, . . ., etc.

(c) By the same reasoning as in (a) and (b), an organism that must code for 64 different traits, would need linear sequences of three nucleotides each to provide 64 possible arrangements (4^3), for example, AAA, ATA, ACA, AGA, AAT, . . ., etc.

Note that each of these answers refers to a linear sequence of nucleotides, that is, to the length of nucleotides present on a single strand of DNA.

4-2. As demonstrated in Problem 4-1, combinations of three nucleotides provide 64 different triplets, which is more than enough information to represent each of the 20 amino acids used in protein synthesis. On the other hand, pairs of two nucleotides would provide only 16 sequences. The triplet is therefore the shortest sequence that codes sufficient information for 20 different amino acid "messages." As will be discussed in Chapter 28, these triplets are called *codons*, and most amino acids are represented by more than one type of codon.

4-3. 3000 (proteins) $\times$ 100 (amino acids each) $\times$ 3 (nucleotides per amino acid) = 900,000 nucleotide length.

4-4. The five different kinds of amino acids in these organisms can be coded by sequences that are two nucleotides long, which would provide more than enough (16) possible different messages (see answer to Problem 4-1b). Thus, for 100 proteins, each 10 amino acids long, the nucleotide length must be 100 ("proteins") $\times$ 10 (amino acids each) $\times$ 2 (nucleotides per amino acid) = 2000.

4-5. (a) DNA, since thymine (T) is present rather than uracil (U).

(b) TAGC

(c) UAGC

4-6. $4^4 = 256$

4-7. If we restrict our attention to the DNA strand that produces the messenger RNA responsible for the synthesis of specific proteins (see text Fig. 4-17), then the complementary DNA strand is obviously different in nucleotide sequence and does not carry the same information. For two complementary DNA strands to code for exactly the same proteins, each nucleotide sequence would have to code for the same amino acid as its complement, for example, TGT would have to code for the same amino acid as ACA. This does not occur. (See also answer to Problem 5-3.)

4-8. More information is necessary since such a ratio says nothing about complementarity of the bases. The bases in the numerator are not complementary to the bases in the denominator, and their ratio may therefore be of any value in both single- and double-stranded DNA.

4-9. In double-stranded DNA, the $(A + C)/(G + T)$ ratio is expected to equal 1 because of the base-pairing between A and T, and C and G [$A = T, C = G; A + C = G + T; (A + C)/(G + T) = 1$]. However, the $(A + T)/(G + C)$ ratio may or may not equal 1 because there is no pairing between the bases in the numerator and those in the denominator (e.g., a double-stranded molecule of DNA may have many AT base pairs and few GC base pairs).

4-10. When base ratios are *not* $A = T(U), G = C$, and $[A + G]/[T (U) + C] = 1$, the nucleic acid is single-stranded. However, when the ratios *are* as given above, the nucleic acid is most probably double-stranded, although such ratios may occur accidentally in some single-stranded forms. For example, species (g) may conceivably have single-stranded DNA in the sequence ATATGGCCGC. It is very unlikely, however, that such ratios occur in single-stranded DNA. The answers are therefore as follows:

(a) DNA, double-stranded   (e) RNA, single-stranded

(b) DNA, double-stranded   (f) DNA, single-stranded

(c) DNA, single-stranded   (g) DNA, double-stranded

(d) RNA, double-stranded

4-11. (a) From the proportions given, the $(A + T)/(G + C)$ ratio can be written as $[A(1) + T(3)]/[G(2) + C(1)] = 1.33 = 4/3$, since $A/T = 1/3$ and $G/C = 2/1$. The $(A + G)/(T + C)$ ratio is then simply $(1 + 2)/(3 + 1) = 3/4$. [Note that *within a single strand* of DNA, the $(A + G)/(T + C)$ ratio is the reciprocal of the $(A + T)/(C + G)$ ratio.]

(b) If we call the original strand *1* and its complement *2*, then *A* in *1* obviously pairs with T in *2*, and G and C act similarly; that is, A/T in strand *1*

= T/A in strand *2*, or the strand *2* A/T ratio is 3, and its G/C ratio is .5. Since the A + G bases in strand *1* pair with the T + C bases in complementary strand *2*, and the T + C bases in strand *1* pair with A + G in strand *2*, we can write the relationship as follows:

strand *1*		strand *2*
$\dfrac{A + G}{T + C} \dfrac{(3)}{(4)}$	=	$\dfrac{T + C}{A + G} \dfrac{(3)}{(4)}$

   The (A + G)/(T + C) ratio in strand *2* is therefore 4/3 = 1.33. In general, the (A + G)/(T + C) ratio in one strand of a double helix is the reciprocal of this ratio in the complementary strand. Note, however, that this reciprocal relationship is not true for (A + T)/(G + C) in the complementary strand of a double helix, because A + T in one strand pairs with A + T in the other, and G + C pairs similarly. Thus the (A + T)/(G + C) ratio is expected to be similar in both strands of a DNA molecule, that is, the complementary strand also has a ratio of 1.33.

   (c) Because of complementary base-pairing, both strands together have ratios A/T = 1; G/C = 1; (A + G)/(T + C) = 1. The (A + T)/(G + C) ratio, however, need not be 1 since there is no pairing between bases in the numerator to those in the denominator (see Problem 4-9). In the present case we know that the (A + T)/(G + C) ratio is 1.33 in each strand, meaning that the ratio for both strands must also be 1.33.

4-12. (a) [A(1) + T(2)]/[G(1) + C(3)] = .75; therefore, [A(1) + G(1)]/[T(2) + C(3)] = .40
   (b) A/T = 2, G/C = 3, (A + G)/(T + C) = 2.5, (A + T)/(G + C) = .75
   (c) A/T = 1, G/C = 1, (A + G)/(T + C) = 1, (A + T)/(G + C) = .75

4-13. Strain A is $T^- L^+$ and strain B is $T^+ L^-$. Transformed B should be $T^+ L^+$, and its presence should be detected on a medium that contains neither thymine nor leucine.

4-14. Transfer of genetic material from one strain to the other would be evidence for conjugation. Therefore, test for the appearance of prototrophs, $T^+ L^+$, by attempting to grow the bacteria on medium containing neither thymine nor leucine.

4-15. Among the various possibilities is to use ultraviolet absorption to help differentiate between nucleic acids (2600 angstroms) and proteins. Feulgen staining would help differentiate between DNA and other materials. Digestion with specific enzymes such as deoxyribonuclease, ribonuclease, and proteases would help determine whether the material was DNA, RNA, or protein, respectively. Some examples of biochemical methods used for this purpose are given by S. Kit (1963, *Ann. Rev. Biochem.*, **32**: 43-82), such as the dissociation of proteins from nucleic acids by sodium perchlorate, the extraction of protein in phenol, and the separation of RNA from DNA by centrifugation in a cesium chloride gradient.

4-16. The number of donor cells providing the transforming DNA is

$$\frac{\text{total DNA used}}{\text{DNA content per cell}} = \frac{10^{-9}}{2 \times 10^{-15}} = 5 \times 10^5.$$

This means that the DNA from 500,000 streptomycin-resistant cells can transform 10,000 streptomycin-sensitive cells; or that only $10,000/500,000 = 2$ percent of the DNA carrying the streptomycin-resistant gene is successful in transforming the recipient cells.

4-17. (a) This statement is incorrect because it is the *sequence* of nucleotide bases in DNA that determines the "message." Two differently appearing organisms must have different nucleotide sequences if their appearances are genetically caused but may nevertheless have similar or even identical base ratios.

(b) This statement is based on a misinterpretation of the function of DNA. DNA is considered to be genetic material because it carries information between cells in the form of nucleotide *sequences*, which are then transcribed and translated into amino acid *sequences* in proteins. The fact that the sequential information in DNA can be changed by outside agents, or that DNA depends on proteins for its reproduction, does not affect its role as the carrier of such sequential information. As far as we know, the translation of *sequential information* goes in only one direction: from DNA to protein.

5
Replication and Synthesis of Nucleic Acids

5-1.   (a)   (b)

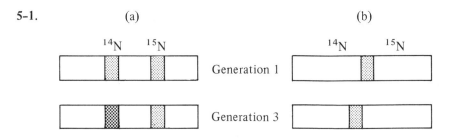

Generation 1

Generation 3

5-2. (a) In a DNA molecule, there is one nucleotide base pair per 3.4 angstroms length (see text p. 46) or, in terms of micrometers, one base pair per 3.4×10^{-4} μm. Therefore the total number of *E. coli* base pairs estimated by this method is: $1100/(3.4 \times 10^{-4}) = 3.24 \times 10^{6}$.

(b) Since there are ten nucleotide base pairs in one complete turn of the DNA double helix, there are $(3.24 \times 10^{6})/10 = 3.24 \times 10^{5}$ turns. If this molecule is assumed to unwind completely in 40 minutes, then the number of revolutions per minute would be $(3.24 \times 10^{5})/40 = 8100$.

5-3. It would matter which strand of DNA was copied from, because each strand would produce a different RNA complement. For example, a DNA strand having the trinucleotide TGT would produce RNA having the sequence ACA, whereas the complementary DNA strand, ACA, would produce RNA having the sequence UGU. According to Table 28-3 (text p. 575), ACA codes for the amino acid serine, and UGU codes for the amino acid cysteine.

5-4. No. Either or both of the two DNA strands would produce RNA with the given base ratio.

5-5. (a) We can assume that the RNA is single-stranded for *B. subtilis* and *E. coli* since the $(A + G)/(U + C)$ ratio departs from 1 in both cases.

(b) No, there is no information on pairing in the $(A + U)/(G + C)$ ratio. For example, a DNA double helix of sequence

$$\begin{array}{c} \text{AAGTAATCTCCA} \\ \text{| | | | | | | | | | | |} \\ \text{TTCATTAGAGGT} \end{array} = \frac{A+T}{G+C} = \frac{16}{8} = 2$$

would produce RNA of ratio $(A + U)/(G + C) = 2$, whether such RNA was formed from both DNA strands or from either one.

   (c) If RNA was copied from both strands of DNA, its total $(A + G)/(U + C)$ ratio would be expected to be 1.00 [similar to the expected $(A + G)/(T + C)$ ratio of a double-stranded molecule of DNA]. Since neither of the $(A + G)/(U + C)$ ratios in these species is 1.00, we can assume they were copied only from single-stranded DNA.

5-6. Because the DNA chain elongates in the $5' \rightarrow 3'$ direction, that is, nucleotides are added at the $3'$ end, type A will act as the inhibitor of DNA synthesis.

5-7. Nearest neighbor frequencies need not be the same since they are not determined by the base ratios but by the order or sequence of bases.

5-8. Organisms with similar nearest neighbor frequencies would have similar base ratios because nearest neighbor frequencies are a good indicator of the frequencies of each base (see Problem 5-11). However, such organisms would not necessarily have the same appearance since the *order* of nearest neighbors may differ between them. For example, TTTTGGGG has fairly similar nearest neighbor frequencies to TTGGGGTT.

5-9. (a) If the two complementary DNA strands had opposite polarity, then $\overset{\text{Ⓒ}}{\underset{T}{\overset{G}{|}}}$ is paired with $\underset{}{\overset{\text{A}}{|}}$ in the following fashion:

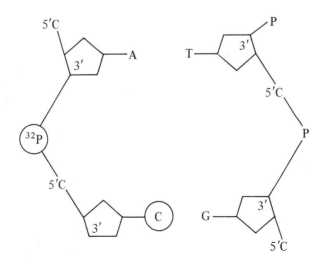

Note that equivalence in nearest neighbor frequencies between $\overset{\textcircled{C}}{\underset{\text{G}}{|}}$ on the "left"

strand and $\overset{\text{G}}{\underset{\text{T}}{|}}$ $\overset{\text{A}}{}$ on its complementary strand would be expected if the labeled

phosphorus atom on the complementary strand were brought in with the thymine nucleotide at its 5' position.* That is, we would expect equal nearest

neighbor frequencies between $\overset{\textcircled{C}}{\underset{\text{A}}{|}}$ and $\overset{\text{G}}{\underset{\textcircled{T}}{|}}$.

    (b) If polarity were the same in the two strands, $\overset{\textcircled{C}}{\underset{\text{A}}{|}}$ is paired with $\overset{\text{G}}{\underset{\text{T}}{|}}$

as follows:

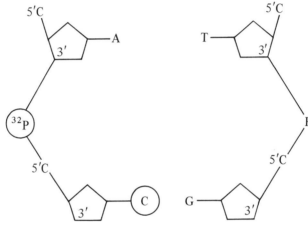

In this case, the ^{32}P label on the strand complementary to that bearing $\overset{\textcircled{C}}{\underset{\text{A}}{|}}$ must

be brought in by the guanine nucleotide, and would then be transferred to the thymine nucleotide by the action of spleen diesterase. Thus, similar polarity

should lead to equivalent nearest neighbor frequencies between $\overset{\textcircled{C}}{\underset{\text{A}}{|}}$ and $\overset{\textcircled{G}}{\underset{\text{T}}{|}}$.

5-10. (a) 1/3 G; 1/3 T; 1/3 C.

    (b) Yes. The labeled G nucleotide on the complementary strand (opposite the C on the illustrated strand) would have its ^{32}P transferred to its nearest neighbor, which would be C. The nearest neighbor frequencies of G in both strands would now be 1/4 G; 1/4 T; 1/2 C.

5-11. To obtain the base ratio for a particular nucleotide, add all of its frequencies as nearest neighbor, for example, thymine = .012 + .026 + .063 + .061 = .162. Thus, the base ratios are .162 thymine, .164 adenine, .337 cytosine, and .337 guanine.

---

*In these studies, the ^{32}P atom is always attached to the carbon at the 5' position of the entering nucleotide, and then, through the action of spleen diesterase, transferred to the 3' position of the neighboring nucleotide.

5-12. (a) If strands are of the same polarity, A3'p5'G = T3'p5'C.

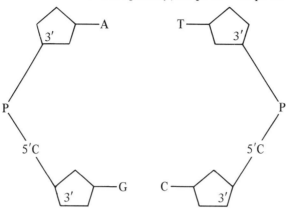

If the two strands have opposite polarity, A3'p5'G = C3'p5'T:

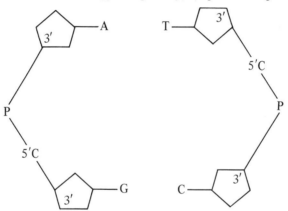

Observed frequencies showed equivalence between ApG and CpT.
(b) TpC
(Reference: E. Chargaff, J. Buchowitz, H. Türler, and H. Shapiro, 1965, *Nature*, **206**:145–147.)

5-13. The absence of G and C in the products indicates that the template DNA possesses neither of these bases. The fact that labeled (A) always has nearest neighbor T and labeled (T) always has nearest neighbor A must therefore mean that the original DNA is an alternating copolymer ATATATAT- - - - - -.

5-14. The information given in Part 1 indicates that all guanine nucleotides are attached together, and they are all at the 5' end of the fragment. If they were placed otherwise, the ^{32}P label at the 5' end of at least one of the guanine nucleotides would become attached, as a result of hydrolysis, to the 3' end of either adenine or cytosine. Since this did not occur, we can tentatively consider the G's to be in the order 5'pGpGpGpG3'. The information in Part 2 indicates that the ^{32}P label from A becomes attached to the 3' end of one of the G's.

Since there is only one 3'G available that can be nearest neighbor to A, the order so far must be 5'pGpGpGpGpA3'. By the same reasoning, the information in Part 3 shows that the complete order of this six-nucleotide sequence must be 5'-GGGGAC-3'.

(Reference: D.R. Mills, F.R. Kramer, and S. Spiegelman, 1973, *Science*, **180**:916-927.)

5-15. (a) Unidirectional:

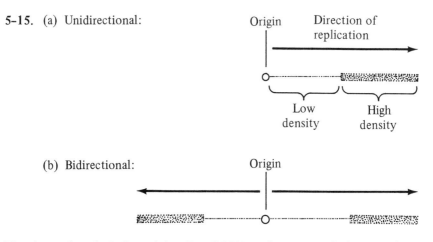

The observed tracks indicated that *E. coli* DNA replication was bidirectional.

(Reference: D.M. Prescott and P.L. Kuempel, 1972, *Proc. Nat. Acad. Sci*, **69**: 2842-2845.)

6

Mendelian Principles:
I. Segregation

6-1. Since *G* (*gray*) is dominant in effect over *g* (*white*), plants that are gray must be either *GG* or *Gg*, and white plants must be *gg*. The presence of white offspring (*gg*) from a test-cross gray (*G?*) × white (*gg*) indicates that the gray plant is a heterozygote and must be carrying the *white* allele (*Gg*). The absence of white offspring from such a cross indicates that the gray parent is a homozygote (*GG*). The appearance of white offspring in one-quarter frequency from a gray × gray cross indicates that both gray parents must be heterozygotes, since only then can offspring receive a *g* allele from each parent. The answers are therefore as follows:

(a) *Gg* × *gg*   (d) *GG* × *gg*
(b) *Gg* × *Gg*   (e) *GG* × *GG* (or *Gg*)
(c) *gg* × *gg*

6-2. In cross (b) two thirds of the gray progeny are expected to be heterozygotes, and would therefore produce white progeny (in one-quarter frequency) when self-fertilized (see text p. 106). In cross (d) all gray progeny are heterozygotes and would be expected to produce one-quarter white progeny. In cross (e) no white progeny would be expected if the initial cross was *GG* × *GG*, and one half of the gray progeny would be expected to produce some white offspring if the cross was *GG* × *Gg*.

6-3. Albinos appear in a frequency of approximately 25 percent among the progeny of these self-fertilized plants. The assumption can therefore be made that albinism is a recessive trait arising, in this case, from self-fertilization of heterozygotes.

(Reference: D.F. Jones, 1915, *J. Hered.*, 6:477–479.)

6-4. (a) If we use the notation *S* = normal, *s* = silky, then *Ss* × *Ss* →
1*SS*:2*Ss*:1*ss*. Phenotypically the progeny are therefore 3/4 normal:1/4 silky, or about $3/4 \times 96 = 72$ normal to $1/4 \times 96 = 24$ silky.

(b) Test-cross it to a silky bird (*S-* × *ss*). If it is heterozygous (*Ss*), at least some silky offspring should appear.

6-5. The observation that closed $\times$ open produces only plants with open flowers indicates that the open trait is dominant to the closed. The further observation that an $F_1 \times F_1$ cross produces about 3/4 open:1/4 closed indicates that the gene determining this trait segregates as a single gene pair, for example, O (open) and o (closed). Similarly, closed $(oo) \times F_1$ (Oo) gives the expected 1 open:1 closed ratio. Thus, the answers are:

(a) oo (b) OO (c) Oo

(Reference: D. Groff and M.L. Odland, 1963, *J. Hered.*, **54**:191–192.)

6-6. (a) The trait can be assumed to be caused by a recessive gene, a, since it appears among the offspring of parents who do not show the trait.

(b) I-1, aa; I-2, AA or Aa; I-3, Aa; I-4, Aa; II-1, Aa; II-2, Aa; II-3, Aa; II-4, AA or Aa; II-5, aa; III-1, aa.

6-7. It appears to be caused by a recessive gene since the trait arises from matings between normal individuals (i.e., normal is dominant) and appears in such instances in a frequency of only 1/4.

6-8. In order for the trait to have appeared (generation V) among progeny of phenotypically normal parents (IV-6, IV-7), we must assume that the parents are heterozygotes. The fact that the initial heterozygous carrier (I-1) is a common ancestor to both IV-6 and IV-7 indicates that transmission of the gene occurred through II-3, who passed it on to III-4 and III-6. Although other individuals in the pedigree *may be* heterozygous, those mentioned above *must be* heterozygous.

6-9. (a) I-1, aa; I-2, Aa; II-1, aa; II-2, Aa; II-3, aa; II-4, Aa; III-1, aa; III-2, Aa; III-3, aa; III-4, Aa; III-5, aa; III-6, Aa; III-7, Aa; III-8, Aa; III-9, aa; III-10, aa.

(b) 1/2
(c) 3/4

6-10. As shown in the accompanying diagram (p. 20), the two different explanations for the appearance of the shaded trait [homozygosity for a recessive allele (a), and heterozygosity for a dominant allele (A)] demand different genotypes for the individuals that "marry into" the affected family: I-2, II-5, III-1, and III-6. For the recessive allele explanation (left-hand side), each of these four phenotypically normal individuals must be heterozygous carriers (Aa) of the a allele; whereas in the dominant explanation (right-hand side), none of these four individuals carry the A allele, and they are instead homozygous aa normals. The answers to questions (a) and (b) therefore depend on the frequency in which carriers of the particular allele causing the trait are expected to be found.

(a) The fact that the trait is frequent in the French population makes it likely that normal-appearing heterozygous carriers (Aa) could be sufficiently frequent that they can marry into the affected family in each generation. In other words, this pedigree does not rule out the possibility that the trait is caused by homozygosity for a recessive allele.

(b) The fact that the trait is very rare in the Asian population makes it highly unlikely that the trait is caused by homozygosity for a recessive allele, and that I-2, II-5, III-1, and III-6 are heterozygotes. (Why should all individuals that marry into this family be carriers of a rare allele?) Thus, it seems most likely that the trait is caused by a dominant allele.

6-11. Three quarters should be brachydactylous since brachydactyly is obviously caused by a dominant gene.

6-12. (a) *Bb*.
      (b) 1/2 brown:1/2 blue.
      (c) 3/4 brown:1/4 blue.

6-13. (a) Sorrel.
      (b) 22.5 sorrel:22.5 black.
      (c) $\chi^2 = .356$, which for 1 degree of freedom has a probability greater than .05 (.50 < probability < .70), and the hypothesis therefore can be accepted.

Figure for Problem 6-10

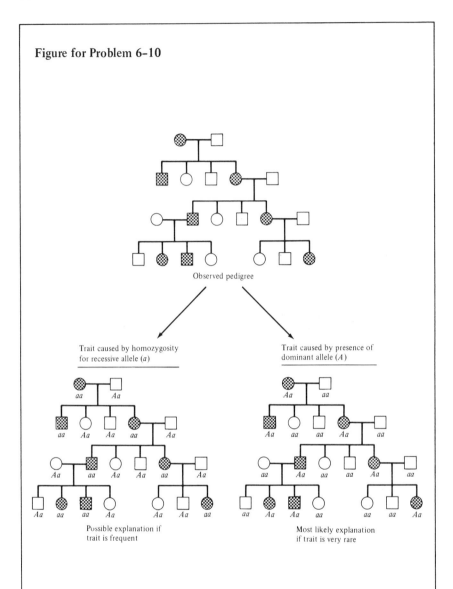

Observed pedigree

Trait caused by homozygosity for recessive allele (*a*)

Trait caused by presence of dominant allele (*A*)

Possible explanation if trait is frequent

Most likely explanation if trait is very rare

Alternative explanations that can be offered for an observed pedigree (*top*). If the frequency of the trait is high, explanation by means of a recessive allele is possible (*left side*). (In that case, additional pedigrees would be needed to discriminate between recessiveness and dominance of the trait.) If the trait is very rare, a recessive cause for the trait in this pedigree is very unlikely, and explanation by means of a dominant allele (*right side*) is much more likely.

7

Mendelian Principles:
II. Independent Assortment

7-1. (a) P_1; *TTSS* $\times$ *ttss* $\rightarrow$ F_1; *TtSs* $\times$ *TtSs* $\rightarrow$ F_2; 9 tall smooth:3 short smooth:3 tall wrinkled:1 short wrinkled.

(b) No. The F_1 is the same: *TTss* $\times$ *ttSS* $\rightarrow$ F_1; *TtSs*.

(c) *TtSs* $\times$ *ttss* $\rightarrow$ 1/4 tall smooth: 1/4 tall wrinkled:1/4 short smooth:1/4 short wrinkled.

7-2. A simple approach to deciphering genotypes when phenotypes are given is to consider that a parent with the dominant phenotype for a particular character may be homozygous or heterozygous, whereas a parent with the recessive phenotype *must be* homozygous for the recessive allele; for example, an albino, short parent is either *ccSS* or *ccSs*. To distinguish between parents carrying the homozygous dominant and heterozygous genotypes, the presence of offspring that show the recessive phenotype for this character would indicate that the parent in question must have been a heterozygote, since homozygous recessive offspring receive a recessive allele from each parent. For example, if a cross of albino short (*ccS-*) $\times$ albino short (*ccS-*) produces offspring that are 3 albino short: 1 albino long, this would mean that both albino short parents are *ccSs* (i.e., *ccSs* $\times$ *ccSs*). Similarly, a cross of dark short (*C-S-*) $\times$ dark long (*C-ss*) that yields offspring in the ratio 3 dark short:3 dark long:1 albino short:1 albino long, means that the parents are *CcSs* $\times$ *Ccss*. The answers are therefore:

(a) *CcSs* $\times$ *CcSs*

(b) *CCSs* $\times$ *CCss* (or *Ccss*)
   (also *CcSs* $\times$ *CCss*)

(c) *CcSS* $\times$ *ccSS* (or *ccSs*)
   (also *CcSs* $\times$ *ccSS*)

(d) *ccSs* $\times$ *ccSs*

(e) *Ccss* $\times$ *Ccss*

(f) *CCSs* $\times$ *CCSs* (or *CcSs*)

(g) *CcSs* $\times$ *Ccss*

7-3. Since the normal alleles for both characteristics are dominant, the crosses can be described as P_1: vg^+/vg^+ e/e $\times$ vg/vg e^+/e^+ $\rightarrow$ F_1: vg^+/vg e^+/e $\times$ vg^+/vg e^+/e. The F_2 phenotypes derive from the following probability calculations:

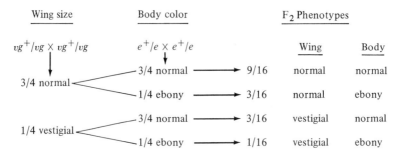

We would therefore expect a 9:3:3:1 ratio among the progeny of the $F_1 \times F_1$ cross; that is, $9/16 \times 512$ normal for both characteristics (288), $3/16 \times 512$ ebony (96), $3/16 \times 512$ vestigial (96), and $1/16 \times 512$ vestigial ebony (32).

7-4. (a) The white and smooth traits appear to be recessive whereas the black and rough traits appear to be dominant since the F_1 is phenotypically both black and rough. White only appears again in the F_2 in a ratio of 90 white:310 black, or approximately 1/4, as would be expected in the F_2 segregation of a trait caused by homozygosity for a recessive allele. Similarly, the F_2 smooths are found in the ratio of 90 smooth:310 rough, or about 1/4. The simplest explanation is therefore that coat color and hair shape in this example are each determined by a separate gene pair, that is, a total of two independently assorting gene pairs are involved.

(b) Color: A (dominant, black), a (recessive, white). Hair: B (dominant, rough), b (recessive, smooth).

(c) $Y = AAbb$ and $Aabb$; Z = aabb.

(d) Expected numbers are those of an F_2 for two independently assorting gene pairs, that is, a 9:3:3:1 ratio. This ratio is then used to calculate chi-square as follows:

	Rough Black	Smooth Black	Rough White	Smooth White	Total
Obs. nos. (o)	240	70	70	20	400
Exp. nos. (e)	$9/16 \times 400 = 225$	$3/16 \times 400 = 75$	$3/16 \times 400 = 75$	$1/16 \times 400 = 25$	400
$o - e$	15	−5	−5	−5	
$(o - e)^2$	225	25	25	25	
$(o - e)^2/e$	$225/225 = 1.000$	$25/75 = .333$	$25/75 = .333$	$25/25 = 1.000$	

Thus, $\chi^2 = 2.67$, which at 3 degrees of freedom has a probability greater than .05 (.30 < probability < .50), and the 9:3:3:1 hypothesis is therefore not rejected.

7-5. Note that a mating of ragged × ragged gives (186 + 174 =) 360 ragged: (57 + 63 =) 120 smooth, or 3 ragged:1 smooth. The allele for ragged can therefore be considered to produce a phenotypic effect dominant to that for smooth (e.g., $R > r$). On the other hand, the pollen results (243 round:237 smooth) indicate that one parent is heterozygous for pollen shape and the other parent is homozygous, but there is no hint in these data as to which is which. Thus, we can arbitrarily designate round pollen as P and angular pollen as p, or vice versa. The following designations provide one set of consistent answers for (a) and (b).

(a) Ragged leaf = R; smooth leaf = r; round pollen = P; angular pollen = p.

(b) Ragged-leaf round-pollen parent = $RrPp$; ragged-leaf angular-pollen parent = $Rrpp$.

(c) Since the cross is hypothesized to be $RrPp \times Rrpp$, the expected phenotypes can be derived as follows:

$Rr \times Rr$	$Pp \times pp$	Progeny
↓	↓	
	—1/2 round-pollen ⟶	3/8 Class I
3/4 ragged-leaf ⟨		
	—1/2 angular-pollen ⟶	3/8 Class II
	—1/2 round-pollen ⟶	1/8 Class III
1/4 smooth-leaf ⟨		
	—1/2 angular-pollen ⟶	1/8 Class IV

Numbers: Class I = 3/8 × 480 = 180; Class II = 3/8 × 480 = 180; Class III = 1/8 × 480 = 60; Class IV = 1/8 × 480 = 60

(d) Chi-square is calculated as follows:

	Classes				Total
	I	II	III	IV	
Obs. (o)	186	174	57	63	480
Exp. (e)	180	180	60	60	480
$o - e$	6	−6	−3	3	
$(o - e)^2$	36	36	9	9	
$(o - e)^2/e$	36/180 = .2	36/180 = .2	9/60 = .15	9/60 = .15	$\chi^2 = .7$

At 3 degrees of freedom, $.7\chi^2$ lies between probabilities .70 and .90 ($> .05$), and the hypothesis described in answer to question (c) is therefore not rejected.

7-6. (a) There are two gene pairs involved in this cross; one gene pair involves resistance to rust race 22, and the other involves resistance to rust race 24. In both gene pairs the allele conferring resistance has an effect dominant to the allele conferring susceptibility. The F_1 is therefore resistant to both races, and the F_2 is susceptible to each rust race in a frequency of 1/4. The expected F_2 ratio would thus be 9/16 resistance to both rust races: 3/16 resistance to one race only: 3/16 resistance to the other race only: 1/4 × 1/4 = 1/16 susceptibility to both races. (Reference: H.H. Flor, 1956, *Adv. Genet.*, **8**:39–54.)

(b) Since the total number of F_2 observed is 194, the expected numbers of F_2 phenotypes would be $9/16 \times 194 = 109.1$ resistant to both rust races 22 and 24; $3/16 \times 194 = 36.4$ resistant only to race 22; $3/16 \times 194 = 36.4$ resistant only to race 24; $1/16 \times 194 = 12.1$ susceptible to both races 22 and 24.

(c) $\chi^2 = 2.55$, which at 3 degrees of freedom has a probability greater than .05 ($.30 <$ probability $< .50$), and the 9:3:3:1 hypothesis is therefore not rejected.

7-7. (a) The F_1 data show that white feathers are dominant to dark, and pea combs are dominant to single combs. Assuming that each of these traits is determined by a single gene pair that assorts independently of the other, the expected F_2 phenotypic ratios would be $9/16$ double dominant:$3/16$ dominant for one trait only:$3/16$ dominant for the other trait only:$1/16$ double recessive. Since 190 chickens were observed in the F_2, their expected numbers would have been $9/16 \times 190 = 106.9$ white pea; $3/16 \times 190 = 35.6$ white single; $3/16 \times 190 = 35.6$ dark pea; $1/16 \times 190 = 11.9$ dark single.

(b) $\chi^2 = 1.55$, which at 3 degrees of freedom has a probability between .50 and .70, and the hypothesis is therefore certainly not rejected at the 5 percent level of significance. (Reference: W. Bateson and R.R. Saunders, 1902, *Reports to the Evolution Committee of the Royal Society*, 1:1-160.)

7-8. Calling the gene pair for feathers, W (dominant, white) and w (recessive, dark), and that for comb shape, P (dominant, pea) and p (recessive, single), the White Leghorn stock can be designated as *WWpp* and the Indian Game Fowl stock as *wwPP*.
 (a) *WwPp* $\times$ *WWpp* $= 1/2$ white pea:$1/2$ white single.
 (b) *WwPp* $\times$ *wwPP* $= 1/2$ white pea:$1/2$ dark pea.
 (c) *WwPp* $\times$ *wwpp* $= 1/4$ white pea:$1/4$ white single:$1/4$ dark pea:$1/4$ dark single.

7-9. (a) $1/4$
 (b) $1/2$
 (c) *AaBb* $\times$ *AaBb* $=$ *Aa* $\times$ *Aa*   *Bb* $\times$ *Bb*

   $1/4\,AA$ $\times$ $1/4\,BB$ $= 1/16\,AABB$

 (d) 0
 (e) *AaBb* $\times$ *AaBb* $=$ *Aa* $\times$ *Aa*   *Bb* $\times$ *Bb*

   $3/4\,A-$ $\times$ $3/4\,B-$ $= 9/16\,A-B-$

 (f) 1 (all)
 (g) *AaBb* $\times$ *AaBB* $=$ *Aa* $\times$ *Aa*   *Bb* $\times$ *BB*

   $1/4\,aa$ $\times$ all $B-$ $= 1/4\,aaB-$

7-10. (a) $3^4 = 81$ (see text Table 7-2, column 4).

(b) Only 1 out of the 81 F_2 genotypes results from the combination of gametes *abcd* $\times$ *abcd*.

(c) Again, the gametic combination *ABCD* $\times$ *ABCD*, provides only 1 out of the 81 F_2 genotypes.

(d) No. The $F_1 \times F_1$ cross would still be the same (*AaBbCcDd* $\times$ *AaBbCcDd*).

7-11. One half, since she would be expected to produce two types of gametes in approximate equal frequency; one type with 24 chromosomes (two No. 21) and one type with 23 (one No. 21).

7-12. (a) Trait *1* is caused by a recessive gene, that is, *a*, since it appears in the progeny of parents who themselves do not show the trait. Trait *2* is caused by a dominant gene, that is, *B*, since it is a *rare* trait and only appears among the progeny of parents who are themselves affected. (see answer to Problem 6-10.)

Trait *1*   Trait *2*

(recessive)   (dominant)

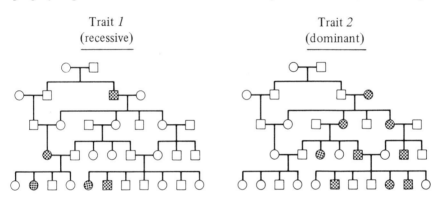

(b) V-5, *aabb*; V-6, *aaBb;* V-7, *AAbb* (or *Aabb*); V-8, *AAbb* (or *Aabb*); V-9, *AABb* (or *AaBb*); V-10, *AABb* (or *AaBb*); V-11, *AAbb* (or *Aabb*).

(c) The mating of V-3 to V-5 can be symbolized *Aabb* $\times$ *aaBb*. The expected progeny are therefore in the following proportions:

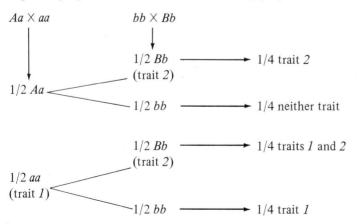

7-13. On the given condition that only first-division segregation occurs for both gene pairs, the tetrads formed should be in the linear order *pe pe pe⁺ pe⁺* (or *pe⁺ pe⁺ pe pe*) as well as *col col col⁺ col⁺* (or *col⁺ col⁺ col col*). Since assortment is independent between the two gene pairs, half the time we would expect *col⁺* to be associated with *pe* (and *col* with *pe⁺*), and half the time *col⁺* should be associated with *pe⁺* (and *col* with *pe*). That is, the tetrads produced should be in the proportions 1/2 parental ditypes (*pe col⁺, pe col⁺, pe⁺ col, pe⁺ col*), and 1/2 nonparental ditypes (*pe col, pe col, pe⁺ col⁺, pe⁺ col⁺*).

7-14. Because of the approximate equivalence of parental ditypes (20) to nonparental ditypes (18), the two genes can be considered to assort independently ($\chi^2_{1\,df} = .026$).

7-15. One half (436) should be of the nonparental ditype. (Reference: C.C. Lindegren, 1933, *Bull. Torrey Bot. Club*, **60**:133–154.)

8

Probability and Statistical Testing

8-1. (a) The different kinds of gametes that can be formed by a heterozygous individual depend only on the number of gene pairs (n) for which the individual is *heterozygous*, and is equal to 2^n. In the present case this is four; therefore $2^n = 2^4 = 16$. (See text Table 7-2, column 2 for the numbers of different kinds of gametes produced by heterozygotes.)

(b) $3^n = 3^4 = 81$. (See text Table 7-2, column 4 for the different kinds of genotypes produced by matings between heterozygotes.)

(c) $1/2$

(d) $1/2 \times 1/4 = 1/8$

(e) A zygote is produced by a chance combination of gametes, each with its own probability. That is, the probability for a child to have a particular genotype is not affected by the genotypes of its siblings. The answer is therefore $1/4$. (Note that this question is different from asking the probability that *a series* of children born to a particular mating will possess given genotypes. In that case, the probability of *a combination* of zygotes is required, and the formulas for such combinations are given on text pp. 136–139.)

8-2. (a) The trait is obviously recessive since it appears among the offspring of parents who themselves do not show the trait. The genotypes are therefore: I-1, *Aa*; I-2, *Aa*; II-4, *aa*; III-2, *Aa*; IV-1, *Aa*; V-1, *aa*.

(b) When affected individuals are homozygous recessive (*aa*), *unaffected* individuals, by definition, cannot be *aa* but must possess one or the other of the remaining two possible genotypes, *AA* or *Aa*. Note, however, that these two genotypes appear in a ratio of 1:2, respectively, among the progeny of a cross between heterozygotes (*Aa × Aa → 1AA:2Aa:1aa*). In other words, two out of three *unaffected* offspring from such a cross would be expected to be heterozygotes. In the present case, the parents of V-1 are obviously heterozygous, and we know that the siblings of V-1 are phenotypically unaffected and must be *AA* or *Aa*. The probability that one of V-1s unaffected siblings would be heterozygous is therefore $2/3$.

(c) $2/3 \times 2/3 = 4/9$

(d) $1/3$, by the same reasoning as (b) above.

(e) From the fact that one of his parents is homozygous, individual V-5 is definitely a heterozygote, and therefore has a 1/2 probability of passing on his recessive gene to his offspring.

(f) 1/2

8-3. (a) 2/3. Since the son is normal, he can be either Aa or AA. As explained above, the ratio of Aa to AA is 2:1, or the probability of Aa in this case is 2/3.

(b) (Probability of parents being heterozygous) $\times$ (probability of affected offspring from mating between heterozygotes) = (2/3 $\times$ 2/3) $\times$ (1/4) = 4/36 = 1/9.

(c) 1/4. The mating is now evidently $Aa \times Aa$.

(d) 3/4

8-4. (a) The physician is in error because the probability for a particular child in this sibship to be affected by this genetic disease (1/4) is not modified by the genotypes of previous children. Each zygotic combination of genes is formed by chance events that are independent of the formation of other combinations.

(b) This defect is obviously caused by a recessive gene, meaning that the normal-appearing parents in generation III are heterozygotes. The chances for IV-1 and IV-6 to produce an affected child then depends on whether they are carrying the gene, that is, whether they are heterozygous. Since the generation III parents produce two heterozygotes among each three normals, the normal-appearing individuals IV-1 and IV-6 must each have a 2/3 chance of being heterozygous. Thus, the probability that they will produce an affected child is the probability that they are heterozygous $\times$ the probability that they will produce an affected child if they are heterozygous or (2/3 $\times$ 2/3) $\times$ (1/4) = 1/9.

(c) If an affected child is produced, this establishes that both IV-1 and IV-6 are heterozygous for this recessive trait. The probability that their next child will be affected is therefore 1/4.

(Reference: A. Freire-Maia, 1971, *J. Hered.*, **62**:53.)

8-5. (a) $1/2 \times 1/2 \times 1/2 = 1/8$

(b) $1 \times 1 \times 1/2 = 1/2$

(c) $Aa \times Aa$ $Bb \times Bb$ $Cc \times Cc$

$\downarrow$ $\downarrow$ $\downarrow$

1/4 AA $\times$ 1/4 BB $\times$ 1/4 CC $=$ 1/64 $AABBCC$

(d) 0, since all zygotes would be $AaBbCc$.

(e) $Aa \times Aa$ $Bb \times Bb$ $CC \times cc$

$\downarrow$ $\downarrow$ $\downarrow$

3/4 $A-$ $\times$ 3/4 $B-$ $\times$ all $C-$ $=$ 9/16 $A-B-C-$

(f) 1

(g) $Aa \times Aa$ $Bb \times BB$ $CC \times cc$

$\downarrow$ $\downarrow$ $\downarrow$

1/4 aa $\times$ all $B-$ $\times$ all $C-$ $=$ 1/4 $aaB-C-$

(h) $1/4 \times 1/4 \times 1/4 = 1/64$

(i) $1/2 \times 1/2 \times 1/4 = 1/16$

(j) 0

8-6. (a) The chances for an F_2 plant to be recessive for all four traits is equal to multiplying the chances that it will be recessive for each, or $1/4 \times 1/4 \times 1/4 \times 1/4 = 1/4^4 = 1/256$.

(b) The chances for an F_2 plant to be phenotypically dominant for each trait is $3/4$. The probability of an F_2 plant dominant for all four traits is therefore $(3/4)^4 = 81/256$.

(c) The proportion of dominant phenotypes among the F_2 equals $(3/4)^4 = 81/256$. The proportion of plants recessive for one or more of the genes is equal to the remainder of F_2 plants, or $1 - (3/4)^4 = 175/256$. The answer is therefore *no*, since $81/256 < 175/256$.

(d) The F_1 parent is heterozygous for all four gene pairs, and the chances to be heterozygous among the F_2 for a particular gene is $1/2$. For all four genes, the answer is therefore $(1/2)^4 = 1/16$.

(e) No. The F_1 plant are still *AaBbCcDd*.

8-7. Using the contingency chi-square test (text pp. 133–134), $\chi^2 = 4.86$. For 1 degree of freedom, this value is significant at the .05 level of significance and indicates that the heterozygotes (Hb^S/Hb^A) in these data suffer fewer heavy infections by the *Plasmodium falciparum* parasite than do the normal homozygotes (Hb^A/Hb^A).

(Reference: A.C. Allison and D.F. Clyde, 1961, *B. Med. J.*, *i*:1346-1349.)

8-8. (a) At the 0.5 level of significance, the sex ratio differs significantly from a 50:50 ratio (100 of each sex expected; $\chi^2_{1df} = 4.20$).

(b) The color ratio does not differ significantly from a 50:50 ratio (100 of each color expected; $\chi^2_{1df} = .40$)

(c) A test for independence gives a χ^2_{1df} of 6.83 (.001 < probability < .01). Color therefore is not independent of sex in this sample.

8-9.

	χ^2	Degrees of Freedom	Probability
Plant 1	.59	1	
Plant 2	1.03	1	
Plant 3	.02	1	
Plant 4	.42	1	
Total	2.06	4	
Summed data	.14	1	.70-.90
Homogeneity	1.92	3	.50-.70

The homogeneity χ^2 shows no significant differences between the plants, and the summed data therefore can be used to test the fit to a 3:1 ratio. As shown above, this χ^2 (0.14) has a probability greater than .05, and the hypothesis of a 3:1 ratio is not rejected.

8-10.

	χ^2	Degrees of Freedom	Probability
Plant A	.22	1	
Plant B	2.97	1	
Plant C	1.08	1	
Plant D	.16	1	
Plant E	1.17	1	
Plant F	1.94	1	
Total	7.54	6	
Summed data	.16	1	.70–.90
Homogeneity	7.38	5	.10–.20

The homogeneity chi-square allows the hypothesis of no significant differences between the plants to be accepted at the 5 percent level of significance (or even at the 10 percent level). The summed data therefore can be used to test the fit to a 3:1 ratio. Since the summed data chi-square of .16 at 1 degree of freedom has a probability much greater than .05, the hypothesis of a 3:1 ratio is not rejected.

8-11. Since this is a backcross to a homozygous recessive stock, the expected ratios for independent assortment are 1 wild:1 black:1 sepia:1 black-sepia. The results of the homogeneity test are

	χ^2	Degrees of Freedom	Probability
Experimenter 1	.49	3	
Experimenter 2	.67	3	
Experimenter 3	.60	3	
Experimenter 4	1.09	3	
Total	2.85	12	
Summed data	1.23	3	.70–.90
Homogeneity	1.62	9	> .95

These data appear to be extremely homogeneous, and there is no question that they can be pooled. For the summed data, χ^2 is well within the bounds of acceptance, and the hypothesis of a 1:1:1:1 ratio is certainly not rejected.

8-12. On the basis of independent assortment, the expected ratios produced by a mating between heterozygotes for black and sepia are 9 wild type:3 black:3 sepia:1 black-sepia. Results of the homogeneity test are

	χ^2	Degrees of Freedom	Probability
Experimenter 1	.62	3	
Experimenter 2	.50	3	
Experimenter 3	.32	3	
Experimenter 4	.33	3	
Total	1.77	12	
Summed data	.14	3	> .95
Homogeneity	1.63	9	> .95

The data therefore can certainly be pooled, and the 9:3:3:1 ratio is not rejected.

8-13. $p = 333/586 = .568; q = 1 - .568 = .432; n = 586.$
   95 percent confidence limits $= .568 \pm 1.96 \sqrt{(.568)(.432)/586} = .568 \pm .040.$

8-14. $.42 \pm .10$

8-15.   (a) $\frac{12!}{12!}(9/16)^{12} (7/16)^0 = (9/16)^{12} = .001$

(b) It might well indicate that homozygous recessive zygotes are not viable; or that, in this case, gametes carrying recessive genes are not viable; or that an error has been made in the mating, and the plant was really fertilized by a plant homozygous for both dominant genes.

8-16.   (a) Note that the question pertains essentially to the types of families rather than the types of children; that is, the question asks, "What is the expected proportion of two-child families recognized by the production of at least one albino child, that has both children albino?" If we call the albino gene a and the normal gene A, the parents of such families can be described as $Aa \times Aa$. Among the matings of this kind that produce only two children, the following are the probabilities that one or both of the children will be albino:
   (i) First child albino, second child normal $= 1/4 \times 3/4 = 3/16$
   (ii) First child normal, second child albino $= 3/4 \times 1/4 = 3/16$
   (iii) First child albino, second child albino $= 1/4 \times 1/4 = 1/16$
   (The remainder of $Aa \times Aa$ matings will produce only normal children and will occur in frequency $3/4 \times 3/4 = 9/16$.)
   If we confine ourselves to only the albino-producing matings designated as i, ii, and iii, above, the question then becomes "What is the proportion of the iii sibship type among i, ii, and iii?" The answer to this is simply $(1/16)/(3/16 + 3/16 + 1/16) = 1/7$.
   (b) In families that fall into categories i and ii in the above answer, the frequency of albino children is $1/2$. Among the three kinds of albino-bearing families that produce sibships of 2, such families occur with probability

$$\frac{(3/16) + 3/16)}{(3/16) + (3/16) + (1/16)} = \frac{6}{7} .$$

Thus albino children occur in families i and ii with relative frequency $1/2 \times 6/7 = 3/7$. However they also occur in iii-type families with relative frequency of $1 \times 1/7 = 1/7$. The total frequency of albino children in two-child families that have at least one albino child is therefore expected to be $3/7 + 1/7 = 4/7 = .5714$.
   A simple formula that can be used to obtain the proportion of albino children in those families that have already been identified as albino-bearing is $(1/4)/[1 - (3/4)^n]$ where n represents the size of the sibship. In the present case, n is 2, and the proportion of albino children in all such families is therefore

.2500/.4375 = .5714. (In general, if q is the probability for the trait to appear among the progeny of particular kinds of matings that have been identified by producing at least one offspring bearing the trait, then the actual proportion of children bearing the trait in such families is $q/[1-(1-q)^n]$.)

8-17. Again, as in Problem 8-16, the matings are of the type $Aa \times Aa$. In five-child sibships produced by such matings, those sibships that contain one or more albino children will have been formed with the following probabilities:

	Probability = $\dfrac{n!}{w!x!}p^w q^x$ (see text p. 136)	Frequency of Given Sibship among Albino-Bearing Families (from Previous Column)	Frequency of Normal Children in Given Sibship
One child albino, four normal	$5(1/4)^1 (3/4)^4 = .3955$	$.3955/.7627 = .5186$	4/5
Two children albino, three normal	$10(1/4)^2 (3/4)^3 = .2637$	$.2637/.7627 = .3457$	3/5
Three children albino, two normal	$10(1/4)^3 (3/4)^2 = .0879$	$.0879/.7627 = .1152$	2/5
Four children albino, one normal	$5(1/4)^4 (3/4)^1 = .0146$	$.0146/.7627 = .0192$	1/5
Five children albino	$(1/4)^5 = .0010$	$.0010/.7627 = .0013$	0
Total probability of an albino–bearing family	.7627	1.0000	

The frequency of normal children in all such families will thus be

$$[(.5186)(4/5)] + [(.3457)(3/5)] + [(.1152)(2/5)] + [(.0192)(1/5)] = .6722.$$

The answer is therefore $.6722 \times 125 = 84$.

If we use the formula described in the latter part of the answer to Problem 8-16b, then the calculations are as follows:

$$\text{proportion albino children} = \frac{q}{1-(1-q)^n} = \frac{1/4}{1-(3/4)^5} = \frac{.2500}{.7627} = .3278$$

The proportion of normal children in these families is therefore $1 - .3278 = .6722$, and the number of such children among 125 progeny is, again, $125 \times .6722 = 84$.

8-18. The 95 percent confidence intervals for the frequencies of tasters and nontasters in each sex can be derived from the table for the binomial distribution (text p. 138):

		95 Percent Confidence Interval
Taster males	$60 = 60/100 = .6$	.50-.70
Taster females	$40 = 40/50 = .8$	.66-.90
Nontaster males	$40 = 40/100 = .4$	.30-.50
Nontaster females	$10 = 10/50 = .20$	.10-.34

Since the confidence intervals for the male and female tasters overlap, there is no clear significant difference between the two sexes in PTC tasting ability.

8-19. (a) $\dfrac{6!}{1!5!}(.67)^1 (.33)^5 = .0157$ (c) $\dfrac{6!}{4!2!}(.67)^4 (.33)^2 = .3292$

(b) $\dfrac{6!}{3!3!}(.67)^3 (.33)^3 = .2162$ (d) $\dfrac{6!}{6!}(.67)^0 (.33)^6 = .0013$

8-20. The sex ratio in this sample is $100\male : 50\female$ or $.67 : .33$.

(a) $\dfrac{3!}{0!3!}(.67)^0 (.33)^3 = .0359$

(b) $\dfrac{3!}{2!1!}(.67)^2 (.33)^1 = .4444$

(c) $1 - \dfrac{3!}{3!0!}(.67)^3 (.33)^0 = 1 - .3008 = .6992$

8-21. (a) $\dfrac{3!}{0!3!}(.50)^0 (.50)^3 = .1250$

(b) $\dfrac{3!}{2!1!}(.50)^2 (.50)^1 = .3750$

(c) $1 - \dfrac{3!}{3!0!}(.50)^3 (.50)^0 = 1 - .1250 = .8750$

8-22. Since the frequency of males is .515, the expected frequency of 12-children families with all boys is $(.515)^{12}$ or .000348, and the expected frequency of all girls in 12-children families is $(.485)^{12} = .000169$. Multiplied by 1,000,000, this gives the expected numbers of all-boy and all-girl families as 348 and 269, respectively. It is apparent that these values differ widely from the observed values of 655 and 56.

(Reference: A. Geissler, 1889. *Z. K. Sachs Statist. Bur.*, **35**:1-24.)

8-23. The mean value is 1 per 10 million bacteria, or, since each dish has 1 million bacteria, we would expect to find an average of one dish in ten with one colony, or $m = 1/10 = .1$. e^{-m} therefore has a value of .905 (text Table 8-8).

	Colonies per Petri Dish	Expected Probability	Expected No. of Dishes
(a)	1	$.905 \times 1 = .905$	$200 \times .905 = 181$
(b)	2	$.905 \times .1 = .0905$	$200 \times .0905 = 18$
(c)	3	$.905 \times (.01)/2 = .0045$	$200 \times .0045 = 1$
(d)	4	$.905 \times (.001)/6 = .0002$	$200 \times .0002 = 0$

8-24. (a) The average mutation rate is 13 colonies per 10 petri dishes of 130 million bacteria each,

$$\frac{13}{10 \times 1.3 \times 10^8} = 1 \text{ per hundred million bacteria } (1 \times 10^8).$$

(b) $m = 13$ colonies per 10 petri dishes $= 1.30$. e^{-m} is therefore .273 (text Table 8-8).

Colonies per Petri Dish	No. Petri Dishes	No. Colonies	Expected Probability	Expected No. of Dishes
0	4	0	$.273 \times 1 = .273$	$10 \times .273 = 2.7$
1	4	4	$.273 \times 1.30 = .355$	$10 \times .355 = 3.5$
2	0	0	$.273 \times \frac{(1.3)^2}{2} = .231$	$10 \times .231 = 2.3$
3	0	0	$.273 \times \frac{(1.3)^3}{6} = .100$	$10 \times .100 = 1.0$
4	1	4	$.273 \times \frac{(1.3)^4}{24} = .032$	$10 \times .032 = .3$
5	1	5	$.273 \times \frac{(1.3)^5}{120} = .008$	$10 \times .008 = .1$

Calculation of χ^2 shows that it is equal to 13.73, which, for 4 degrees of freedom, has a probability somewhat less than .01. This chi-square is certainly significant, although it should be noted that it is caused primarily by the presence of one petri dish in the five-colony category. If a larger number of petri dishes had been used, a better fit to the expected Poisson distribution would undoubtedly have been observed.

(Reference: M.G. Curcho, 1948, *J. Bacterio.*, **56**:374-375.)

9

Dominance Relations and Multiple Alleles in Diploid Organisms

9-1. Since all three phenotypes are produced among their progeny, the parents must be heterozygous; that is, both are pink.

9-2. (a) *RR* × *RR* → all red.
    (b) *RR* × *Rr* → 1/2 red:1/2 pink.
    (c) *Rr* × *rr* → 1/2 pink:1/2 white.
    (d) *Rr* × *Rr* → 1/4 red:1/2 pink:1/4 white.

9-3. (a) *RrHh* = all roan hornless.
    (b) F_1 × F_2: *RrHh* × *RrHh*

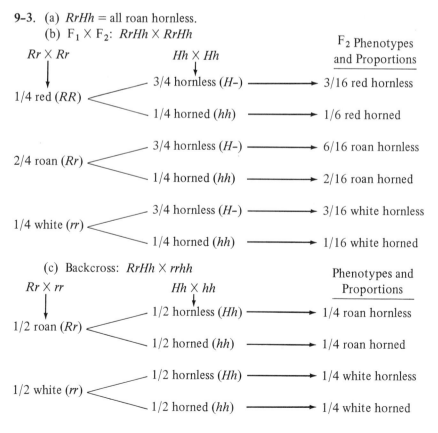

(c) Backcross: *RrHh* × *rrhh*

9-4. (a) Note that crosses of light × light always produce light offspring, and ring × ring always produce ring offspring. On the other hand, light × ring produce a new phenotype, buff, in all of the progeny. It is likely therefore that light and ring are homozygotes, and buff is the heterozygote bearing both a light and ring allele. This view is supported by the fact that crosses of buff × buff produce the three phenotypes, light, buff, and ring, in the proportion 1:2:1 as would be expected in a cross between two heterozygotes, $A^1A^2 \times A^1A^2 \rightarrow 1\,A^1A^1{:}2\,A^1A^2{:}1\,A^2A^2$. Thus the crosses can be described as illustrating the segregation of two alleles of a single gene pair in which the heterozygote (A^1A^2) shows incomplete dominance, or a somewhat different phenotype than either homozygous parent.

(b) $A^1A^1 = $ light; $A^1A^2 = $ buff; $A^2A^2 = $ ring.

(Reference: V.S. Asmundsen, U.K. Abbott, and F.H. Lantz, 1964, *J. Hered.*, **55**:150-153.)

9-5. The allele producing dry earwax seems to be recessive since dry × dry crosses produce only dry offspring, indicating that dry individuals must be homozygous. Sticky-earwax individuals, on the other hand, can apparently be heterozygous, since sticky × sticky produce some dry offspring. The fact that such parents appear as sticky indicates that sticky is dominant in the heterozygote.

(Reference: N.L. Petrakis, K.L. Molohon, and D.J. Tepper, 1967, *Science*, **158**:1192-1193.)

9-6. (a) 5 different possible phenotypes.

(b) 7

(c) 9

One way to determine the number of different possible phenotypes when dealing with independently segregating pairs of genes affecting the same trait and in which dominance is absent is to use the binomial expansion. For example, in an individual heterozygous for n pairs of genes determining skin color, each pair having one black and one white allele, there are (1/2 black + 1/2 white)^n kinds of gametes possible, and (1/2 black + 1/2 white)^n × (1/2 black + 1/2 white)^n = (1/2 black + 1/2 white)^{2n} possible zygotic combinations, which are distributed according to the coefficients in text Table 8-6. For two pairs of genes, the binomial is therefore expanded to 4, for three pairs to 6, etc. One then looks up the number of terms in text Table 8-6 corresponding to n = 4, n = 6, etc., and these are, as given above, five phenotypes, seven phenotypes, etc. (see also text p. 246).

9-7. (a) $CC \times c^k c^k$ → all black (Cc^k)

(b) $CC \times c^d c^d$ → all black (Cc^d)

(c) $CC \times c^a c^a$ → all black (Cc^a)

(d) $c^k c^k \times c^d c^d$ → all sepia $(c^k c^d)$

(e) $Cc^k \times Cc^a$ → 3 black $(C\text{-})$:1 sepia $(c^k c^a)$

(f) $Cc^k \times c^k c^d$ → 1 black $(C\text{-})$:1 sepia $(c^k\text{-})$

(g) $Cc^d \times c^k c^d$ → 2 black $(C\text{-})$:1 sepia $(c^k c^d)$: 1 cream $(c^d c^d)$

9-8. (a) $Cc^a \times Cc^a$
    (b) $Cc^k \times c^a c^a$
    (c) $c^d c^a \times c^d c^a$
    (d) $c^k c^a \times c^d c^a$
    (e) $Cc^d \times c^a c^a$

    (f) $Cc^k \times c^d c^d$
    (g) $Cc^k \times c^k c^k$
    (h) $Cc^d \times c^k c^d$
    (i) $c^k c^d \times c^k c^d$
    (j) $c^d c^a \times c^a c^a$

9-9. If the sepia parent were homozygous, 50 percent of the offspring would be expected to be sepia. The 95 percent confidence interval for an observed number of 6 out of 20 in a binomial distribution (text Table 8-7) is from 12 to 54 percent. Thus, the observed ratio does not exclude the possibility of a 50 percent frequency, and the sepia parent may well have been homozygous for the sepia gene. (A chi-square test gives $\chi^2_{1df} = 2.44$. The hypothesis of a sepia homozygous parent is therefore not rejected.)

9-10. (a) Palomino is apparently an incompletely dominant effect produced in a heterozygote for a single gene difference, for example, $A^1 A^1$ = chestnut, $A^1 A^2$ = palomino, $A^2 A^2$ = cream.

    (b) According to this hypothesis, matings between palominos should produce 1 chestnut:2 palomino:1 cream, or, among 84 offspring, 21 chestnut:42 palomino:21 cream. Chi-square is .29 which, at 2 degrees of freedom, has a very high probability (.70-.90), and the hypothesis is therefore not rejected.

    (c) Mate palominos with chestnuts.

9-11. The simplest hypothesis is that there is only one pair of genes segregating in these crosses, with three different possible alleles: *red*, *colorless*, and *variegated*. The effect of *red* appears to be dominant to the effects of the other two alleles, and the heterozygote from a cross between the red strain and either of the other two strains is always phenotypically red. It is only in the F_2 that the recessive phenotype reappears, but then in ratios of one quarter, which indicates that both *colorless* and *variegated* are allelic to *red*; that is, *colorless* and *variegated* are allelic to each other.
(Reference: R.A. Emerson, 1911, *Reports Nebraska Agric. Exp. Sta.*, 24:58-90.)

9-12. The gene producing hemoglobin C seems to be allelic to the genes producing hemoglobins A and S since, in each individual in which hemoglobin C appears, only one other hemoglobin appears, either A or S, but not all three together. In order words, C appears to be inherited as an alternate of hemoglobins A and S in a multiple allelic system. This view has been supported by considerable data gathered since these pedigrees were discovered (see also text pp. 535ff.).
(Reference: H.A. Itano and J.V. Neel, 1950, *Proc. Nat. Acad. Sci.*, 36:613-617.)

9-13. These plumage patterns are evidently caused by multiple alleles of a single gene with dominance relationships as follows: the effect of the *restricted* allele P^R is dominant over the *mallard* allele P^M (cross 1) and over the *dusky* allele P^D (cross 3), and the effect of the *mallard* allele is dominant over *dusky*.

(a) F_1 of cross 1 is restricted in phenotype ($P^R P^M$), and the F_1 of cross 2 is mallard ($P^M P^D$). A cross between these F_1 produces genotypes in the ratio 1 $P^R P^M$:1 $P^R P^D$:1 $P^M P^M$: 1 $P^M P^D$, or phenotypes in the ratio 1 restricted:1 mallard.

(b) A cross $P^R P^D \times P^M P^D$ produces genotypes in the ratio 1 $P^R P^M$:1 $P^R P^D$:1 $P^M P^D$:1 $P^D P^D$, or phenotypes in the ratio 2 restricted:1 mallard:1 dusky.

9-14. The fact that the trait is rare in such populations makes it highly unlikely that wooly haired individuals will be found in sufficient frequency to mate with each other. That is, it is safe to assume that the wooly haired mother is a heterozygote, that is, one of her parents was straight-haired.

(a) $1/2 \times 1/2 = 1/4$
(b) $1/2 \times 1/2 = 1/4$
(c) $1/4$

9-15. (a) The direct transmission of the rare V trait between generations II and III, and the fact that it appears phenotypically in II–2 and three of her offspring, indicates that it is probably caused by a dominant gene.

(b) It could be an allele of the MN gene since the offspring of female II–2 that show trait V also have not inherited their mother's N allele. Similarly, her offspring that show the maternal N gene also have not inherited the maternal V trait. In other words, it appears as though female II–2 is heterozygous N/V, and children who receive her V allele do not get her N allele at the same time, nor do children receiving her N allele get her V allele at the same time. This makes V a codominant allele with M and N. (Note also that if the generation II cross were of genotypes $MM \times NN$, and V were *not* an allele of M and N, we would expect all offspring to be of MN phenotype. This is obviously not so.) (Reference: H. Gershowitz and K. Fried, 1966, *Am. J. Hum. Genet.* **18**:264–281.)

9-16. Since all individuals in this sample react with one or the other of the two sera, we can assume that each individual possesses either one (homozygotes) or both (heterozygotes) of the two antigens that have been tested. That is, according to the simplest reasoning, only one gene pair with two alleles can explain these data. Thus, if we designate the allele for the antigen precipitated by anti-α as A^α, and the anti-β antigen allele as A^β, then class 1 is genotypically $A^\alpha A^\beta$, class 2 is $A^\alpha A^\alpha$, and class 3 is $A^\beta A^\beta$.

9-17. The γ antiserum probably recognizes another allele at this gene pair (e.g., A^γ), so that about half of the individuals in class 2 are $A^\alpha A^\gamma$ (and the other half are presumably $A^\alpha A^\alpha$), whereas half of those in class 3 are $A^\beta A^\gamma$ (and the other half of class 3 are $A^\beta A^\beta$). The conclusion that A^γ is a third allele of the A gene is reinforced by the observation that none of the individuals in class 1 ($A^\alpha A^\beta$ genotype) react with γ antiserum, that is, there is only room in any one individual for two alleles of a gene pair. [The absence of individuals who react only with antiserum γ can be explained by the restriction of classes 1, 2, and 3, to individuals who react with antiserums α and/or β. Presumably there are other individuals in the population ($A^\gamma A^\gamma$) who react only with antiserum γ.]

9-18. Only two of the parental phenotypes, AB and O, can be assigned complete genotypes (*AB, OO*) based on their phenotypes alone. The A and B parents may be homozygous (*AA, BB*) or heterozygous (*AO, BO*), depending on the appearance of their offspring. The appearance of O offspring from an A or B parent certainly indicates that the parent was a heterozygote carrying the recessive allele *O*. Similarly, the appearance of A or B offspring from a cross of A × B parents indicates that one or both parents are heterozygous, since such offspring can only be heterozygous (AO or BO) from such a cross. In crosses of the type A × AB or B × AB, the appearance of B and A offspring, respectively, also indicates that these are heterozygous, and the A or B parent must be heterozygous. (For example, the appearance of B offspring from an A × AB cross means that the A parent has contributed an *O* allele and the AB parent has contributed the *B* allele.) The answers are therefore as follows:

(a) *BO* × *BO*   (e) *BO* × *AO*
(b) *BB* × *AB*   (f) *BO* × *AB*
(c) *BB* × *AO*   (g) *BB* × *OO*
(d) *BO* × *AA*   (h) *BO* × *OO*

9-19. (a) Individuals 1 and 6 belong to group O since their blood cells ("no antigens") are accepted by all the others, but their serum (antibodies to A and B) agglutinates all the other blood cells.

(b) *AB* genotypes would be expected to accept *all* blood cells since they have no ABO antibodies. None of the six individuals are in this category.

(c) $1 = O, 2 = A, 3 = B, 4 = B, 5 = A, 6 = O$

(Reference: K. Landsteiner, 1901, *Wien. Klin. Wochenschr.*, **14**:1132-1134. This paper has been translated into English and is reprinted in Boyer's collection: see References in text Chapter 1.)

9-20. (a) All A   (e) 3/4 A:1/4 O
(b) 1/2 A:1/2 AB   (f) 1/2 A:1/4 AB:1/4B
(c) 1/2 A:1/2 AB   (g) 1/2 A:1/2 O
(d) All A

9-21. (a) $I^A i\, Rr \times I^B I^B\, rr$

(b) *iiRr* × *I^A irr*

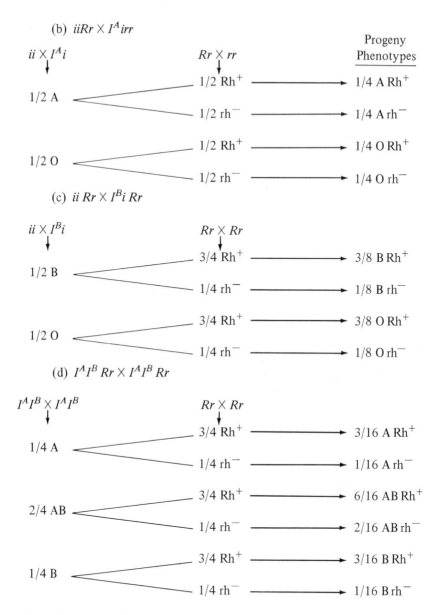

ii × *I^A i*   *Rr* × *rr*

Progeny
Phenotypes

1/2 A
   1/2 Rh⁺ ⟶ 1/4 A Rh⁺
   1/2 rh⁻ ⟶ 1/4 A rh⁻

1/2 O
   1/2 Rh⁺ ⟶ 1/4 O Rh⁺
   1/2 rh⁻ ⟶ 1/4 O rh⁻

(c) *ii Rr* × *I^B i Rr*

ii × *I^B i*   *Rr* × *Rr*

1/2 B
   3/4 Rh⁺ ⟶ 3/8 B Rh⁺
   1/4 rh⁻ ⟶ 1/8 B rh⁻

1/2 O
   3/4 Rh⁺ ⟶ 3/8 O Rh⁺
   1/4 rh⁻ ⟶ 1/8 O rh⁻

(d) *I^A I^B Rr* × *I^A I^B Rr*

I^A I^B × *I^A I^B*   *Rr* × *Rr*

1/4 A
   3/4 Rh⁺ ⟶ 3/16 A Rh⁺
   1/4 rh⁻ ⟶ 1/16 A rh⁻

2/4 AB
   3/4 Rh⁺ ⟶ 6/16 AB Rh⁺
   1/4 rh⁻ ⟶ 2/16 AB rh⁻

1/4 B
   3/4 Rh⁺ ⟶ 3/16 B Rh⁺
   1/4 rh⁻ ⟶ 1/16 B rh⁻

9-22. The rationale for determining *ABO* parental genotypes is given in the answer to Problem 9-18. The determination of parental *Rh* genotypes can be made on the basis that rh⁻ phenotypes must be homozygous *rr* since this trait is recessive, whereas Rh⁺ phenotypes are heterozygous *Rr* if they produce *rh⁻* offspring in crosses with either Rh⁺(*Rr*) or rh⁻(*rr*) individuals. The answers are therefore

(a) *I^A I^B Rr* × *ii Rr*
(b) *I^A i RR* × *I^A irr*
(c) *I^B i RR* × *I^A i rr*

(d) *I^B i Rr* × *I^A i rr*
(e) *I^B i Rr* × *I^A i Rr*

9-23. The appearance of B blood type in the child means that the father must have contributed a *B* allele. (The mother is obviously a heterozygote and contributed an *O* allele, since if she were homozygous *AA*, the child would be AB.) The N blood type of the child is that of a homozygote *NN* and indicates that the father also contributed an *N* allele. The fact that the child is Rh$^+$, and the mother is rh$^-$, indicated that the *R* allele in the child comes from the father. The possible paternal phenotypes are, therefore, B MN Rh$^+$, B N Rh$^+$, AB MN Rh$^+$, or AB N Rh$^+$.

9-24. Since the offspring has O blood type, it is homozygous *OO*, and the father must have contributed an *O* allele. Similarly, the father must have contributed an *R* allele since the offspring is Rh$^+$ and the mother is rh$^-$ (*rr*). The MN blood type in the child and MN blood type in the mother means that the paternal genotype is not restricted, and may have been either *MM*, *NN*, or *MN*. On this basis we can exclude male phenotypes AB Rh$^+$ M, B rh$^-$ MN, O rh$^-$ N. The only male of this group that is not excluded is A Rh$^+$ MN.

9-25. The genotypes of the parents are determined by examining the two alleles for each gene pair among the offspring and assigning one allele to one parent and one to the other. For example, each parent must be carrying an *i* allele because of the presence of *ii* offspring. Since the offspring genotypes show that the parents are also carrying *I^A* and *I^B* alleles, one parent must be *I^Ai* and the other *I^Bi*. Following this line of reasoning for the other gene pairs, the parental genotypes are $I^A i L^M L^N Rr$, $I^B i L^M L^N Rr$.

9-26. (a) $I^A i\, rr\, L^M L^M \times I^B I^B\, Rr\, L^N L^N$

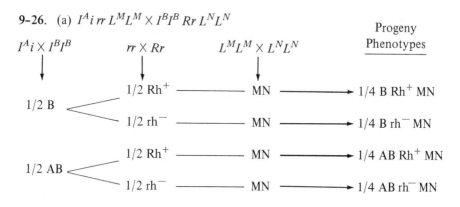

(b) $I^A I^B\, rr\, L^N L^N \times I^B i\, Rr\, L^M L^M$

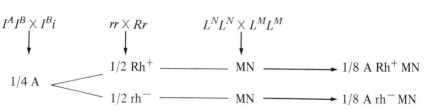

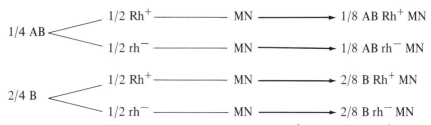

(c) Yes, since the offspring may be of genotype $I^A i$, inheriting its I^A gene from its (a) mother and i gene from the (b) father. The other blood groups would be important to help decide this, that is, if the disputed child of the (a) female was also MN, this would exclude the possibility of a (b) male as the father.

(d) The (a) matings would be more protective since the mother produces B antibodies.

9-27. (a) 4 (Male number 1 could also fit here, but is the only choice for mating with female e.)

(b) 3   (d) 2

(c) 5   (e) 1

9-28. Three different self-sterility alleles are sufficient to explain the results. The fact that type A pollen can produce type A plants when mated to type B females indicates that A and B must share an allele in common, for example, S^1, although the successful A pollen must contain a different allele, for example, S^2. Thus, we can give type A the genotype $S^1 S^2$ and type B the genotype $S^1 S^3$. Since only the S^2 pollen is successful in this cross, two types of plants are produced, $S^1 S^2$ and $S^2 S^3$; the former is type A and the latter can be called type C. Note that these genotypes now account for the observed data.

9-29. (a) Since the F_1 in text p. 156 is heterozygous for three gene pairs, an F_2 individual who accepts F_1 donor tissues must also be heterozygous for the same three pairs. The proportion of the F_2 that is heterozygous for any particular gene pair segregating in an $F_1 \times F_1$ cross is 1/2; therefore the proportion of F_2 heterozygous for three gene pairs is $(1/2)^3 = 1/8$.

(b) To accept tissues from a parental strain, an individual must be carrying the particular parental strain allele either as a heterozygote or as a homozygote. For each particular parental allele, one half of the F_2 is heterozygous and one quarter of the F_2 is homozygous, or $1/2 + 1/4 = 3/4$ of the F_2 bear that allele. Thus, the proportion of the F_2 that carries *all of the three* alleles from one parental strain is $(3/4)^3 = 27/64$.

9-30. (a) $(1/2)^4 = 1/16$

(b) $(3/4)^4 = 81/256$

9-31. 1/4 (*A2 B12/A9 B5*)

1/4 (*A2 B12/A3 B7*)

1/4 (*A1 B8/A9 B5*)

1/4 (*A1 B8/A3 B7*)

10
Environmental Effects and Gene Expression

10-1. (a) The fact that penetrance is not complete will cause a reduction in the frequency of expected camptodactylous phenotypes. The answer is therefore 1/2 (frequency of dominant genotype) × 3/4 (penetrance) = 3/8 camptodactylous phenotypes.

(b) Under the assumption that the reduction in penetrance applies equally to homozygotes and heterozygotes for camptodactyly, the answer is 3/4 (frequency of dominant homozygotes and heterozygotes) × 3/4 (penetrance) = 9/16.

10-2. (a) $1/2 \times 1/4 = 1/8$
(b) $3/4 \times 1/4 = 3/16$

10-3. (a) If penetrance were complete, half the progeny of each affected person would be expected to show the trait. Since penetrance is only 50 percent, $1/2 \times 1/2$, or one quarter of the progeny of affected individuals might be expected to show the trait. Thus, individuals II-4, III-5, and III-7 might be arbitrarily designated to show the trait, or individuals II-2 and III-2. Among the other offspring who do not show the trait, one third might be expected to be heterozygous Aa, since one half of the Aa heterozygotes do not show the trait:

$Aa \times aa$	Appearance of Trait	Phenotypic Ratio
1/2 Aa	1/2 show the trait	= 1/4 (Aa) show the trait
	1/2 do not show the trait → 1/4 (Aa)	
1/2 aa	all do not show the trait → 2/4 (aa)	= 3/4 do not show the trait

Although it is optional which individuals are designated to be heterozygous Aa, there are the restrictions that their parents be carrying the A gene and that the frequency of heterozygotes should be about 50 percent, half of which are affected.

(b) II-1 = aa, II-2 = Aa. (The rarity of the trait would indicate that individual II-2 rather than II-1 was the carrier of the gene. That is, it is unlikely that the trait was brought in secondarily from outside the family that initially showed it.)

10-4. (a) 1/2 (probability that father of the child inherited the gene) × 1/2 (probability that the child received the gene) = 1/4.

(b) Since one of the children shows symptoms of the disease, we now know that the father of the two children inherited the Huntington's chorea gene. The probability that the second child is carrying this gene is therefore now 1/2.

(c) $\dfrac{n!}{w!x!} p^w q^x = \dfrac{2!}{1!1!} \, 3/4 \cdot 1/4 = 6/16$

This same probability can be derived by considering that the probability that one of the two children will carry the gene = 1 − [probability both carry it + probability both do not carry it] = 1 − [(1/4)(1/4) + (3/4)(3/4)] = 1 − 10/16 = 6/16.

10-5. The treated female phenylketonuric is only a phenocopy of the normal (see text p. 168) but is nevertheless still unable to metabolize phenylalanine. Since the adult brain is relatively insensitive to excess amounts of phenylalanine, she may in fact eat a normal phenylalanine-laden diet without showing any symptoms of the disease. However, should such an individual continue eating a normal diet during her pregnancy, relatively high amounts of phenylalanine can cross the placental barrier and seriously affect the mental development of the embryo. Thus, ironically, a genetic mutant who is a "normal" phenocopy can produce a "mutant" phenocopy who is genetically normal.

10-6. A chi-square test for independence yields $\chi^2 = 12.49$, indicating that there is a significant effect of temperature on the penetrance of the trait: as temperature increases from 33°C to 35°C, the frequency of the *scar* phenotype increases significantly.

(Reference: A.E. Bell, D.M. Shideler, and H.L. Eddleman, 1964, *Tribol. Inf. Bull.,* 7:46–48.)

10-7. (a) The data given in the problem show that by 8 to 10 months, 141 out of 337 *hr/hr* homozygotes were leukemic, and at 18 months 243 of these mice were leukemic. On the other hand, only 4 out of 279 *hr/hr$^+$* heterozygotes were leukemic at 8 to 10 months and 56 were leukemic at 18 months. One method of analysis that can be used is to test the incidence of leukemia versus nonleukemia for each genotype irrespective of the age at which it develops, and then to test whether there is a statistical difference between the two genotypes in the age at which leukemia develops. Both of these tests can be performed by calculating contingency chi-squares (see text pp. 133–134):

Incidence of Leukemia

		Leukemic	Nonleukemic	Totals
Genotypes	*hr/hr*	243	94	337
	hr/hr$^+$	56	223	279
		299	317	616

$$\chi^2 = \frac{[|(243)(223) - (94)(56)| - (1/2)616]^2\ 616}{(337)(299)(279)(317)} = 163.8$$

Age of Leukemia Development

Genotypes		Up to 8–10 Months	From about 10 Months to 18 Months	Totals
	hr/hr	141	102	243
	hr/hr^+	4	52	56
		145	154	299

$$\chi^2 = \frac{[|(141)(52) - (102)(4)| - (1/2)299]^2\ 299}{(243)(145)(56)(154)} = 45.16$$

These results show that leukemia occurs among the two genotypes in significantly different ways: greater incidence of leukemia as well as earlier development of leukemia takes place in the hr/hr genotype.

(b) Among hr/hr homozygotes the incidence of leukemia by 8 to 10 months is 141/337 or .418, whereas the same incidence of leukemia among hr/hr^+ heterozygotes is 4/279 or .014. Since a cross between heterozygotes produces 1/4 hr/hr homozygotes and 1/2 hr/hr^+ heterozygotes, the proportion of offspring affected by leukemia at 8 to 10 months of age would be $(1/4 \times .418) + (1/2 \times .014) = .111$.

(Reference: H.D. Meier, D.D. Myers, and R.J. Huebner, 1969, *Proc. Nat. Acad. Sci.,* **63**:759–766.)

10-8. Thorax pigment pattern is obviously a sex-limited trait, so that only females show phenotypic differences. Consequently only information of female phenotypes can furnish genotypic information. The 3:1 ratio produced by female "H" in cross I indicates that only a single gene may be responsible for these thorax patterns, and that the heteromorphic allele is dominant, and the andromorphic allele is recessive in females, for example, $HH = Hh =$ heteromorphic; $hh =$ andromorphic. On this basis, cross I can be considered as $Hh \male \times Hh \female \rightarrow 3H\text{--}:1\ hh$. Since cross II does not produce andromorphic female offspring although mated to the same male as was the "H" female in cross I, the cross II female can be considered homozygous heteromorphic (HH): $Hh \male \times HH \female \rightarrow$ all $H\text{--}$. Cross III must then be of the type $Hh \male \times hh \female \rightarrow 1Hh:1hh$.

(Reference: C. Johnson, 1964, *Genetics,* **49**:513–519.)

10-9. Information that enables a distinction to be made between parental genotypes for this character can be derived only from the different phenotypes of males, since the females are all phenotypically identical. Among males, note that the cock-feathered phenotype must be caused by an allele

(F) that is dominant over that for hen–feathered (f) since the F_1 males of cross (a) are 3/4 cock-feathered:1/4 hen-feathered, as would be expected of progeny from a cross between two heterozygotes. Note also that the appearance of cock-feathering in male progeny when the male parents are hen-feathered in crosses (b) and (c) means that the cock-feathered allele (F) must have been maternally transmitted. The parental genotypes of these crosses can therefore be written as follows:

(a) $Ff\delta \times Ff♀$ (d) $Ff\delta \times ff♀$
(b) $ff\delta \times FF♀$ (e) $FF \delta \times ff$, Ff or $FF♀$;
(c) $ff\delta \times Ff♀$ or $Ff \delta \times FF♀$

10-10. Based on answers to (a) and (d) in the previous problem, there are two possible hypotheses. (1) The parents are $Ff\delta \times Ff♀$, and the expected proportions of male offspring are 3/4 cock–feathered and 1/4 hen–feathered. (2) The parents are $Ff \delta \times ff♀$, and the expected proportions of male offspring are 1/2 cock-feathered and 1/2 hen-feathered. Chi-square for hypothesis 1 is 6.31 and therefore can be rejected at the 5 percent level of significance (1 degree of freedom). Chi–square for hypothesis 2 is 1.50; this hypothesis is therefore not rejected.

10-11. In this case, butterfly color acts similarly to feathering in poultry in the previous problems; that is, only the male phenotypes can be used to distinguish genotypes. Cross (a) shows that the allele causing yellow (C) has an effect that is dominant over white (c) in males, since the F_1 males appear in a ratio of 3 yellow:1 white, as expected from a cross between heterozygotes ($Cc \times Cc$). The appearance of yellow male offspring in crosses (b) and (c) when the male parent is white (cc) indicates that the female parent must have been carrying the yellow allele: in (b) she is Cc, and the cross is $cc \times Cc$; in (c) she must be CC, and the cross is $cc \times CC$. Data on the gene pair for antennae shape indicate that the allele for full antennae (A) is dominant over that for club (a), since a full $\times$ full cross (b) produced 3/4 full:1/4 club, as though the cross were $Aa \times Aa$, and a full $\times$ club cross (c) can produce all full offspring as though the cross were $AA \times aa$. The answers are therefore as follows:

(a) $Ccaa\delta \times CcAa♀$
(b) $ccAa\delta \times CcAa♀$
(c) $ccAA\delta \times CCaa♀$

10-12. One pair of genes appears to be involved in determining this trait, and alleles can be designated H^1 = horned and H^2 = hornless. According to this designation, the Dorset horned breed is H^1/H^1, and the Suffolk hornless breed is H^2/H^2. Note that the data show that H^1 is dominant in males (the F_1 males are all horned), that is, the heterozygote H^1/H^2 and homozygote H^1/H^1 are horned. In females, however, H^1 (horned) is recessive, whereas H^2 (hornless) is dominant, and therefore only the homozygote H^1/H^1 is horned.

10-13. (a) $H^1/H^2 \delta \times H^2/H^2 \circ \to 1H^1/H^2:1H^2/H^2$. The females are therefore all hornless (H^2-), and the males are 1/2 horned:1/2 hornless.

(b) $H^1/H^2 \delta \times H^1/H^1 \circ \to 1H^1/H^1:1H^1/H^2$. The females are therefore 1/2 horned:1/2 hornless, and the males are all horned (H^1-).

10-14. Using the analytical technique described in the text, the trait appears to be almost entirely caused by genetic factors since monozygotic concordance is high and dizygotic concordance is low. There appears to be very little, if any, environmental effect since monozygotic discordance is absent.

(Reference: E.S. Vesell and J.G. Page, 1968, *Science*, **161**:72–73.)

10-15. (a) Heredity seems to play a significant role in determining this trait, since there is such a wide difference between monozygotic and dizygotic concordances. Environmental differences, however, still affect the trait as evidenced by the fact that only about 50 percent of the monozygotic twins are concordant.

(b) In this case, heredity plays a highly significant role, and environmental differences have relatively less effect.

(c) The penetrance of the genes determining trait A is markedly lower (i.e., more environmentally modified) than the penetrance of the genes determining trait B.

10-16. Chi-square tests for independence show a nonsignificant χ^2 for smoking ($\chi^2 = 2.10$) but a highly significant χ^2 for coffee consumption ($\chi^2 = 17.14$). Even without χ^2, the high monozygotic concordance for coffee consumption compared to the relatively low dizygotic concordance indicates an important role of heredity in the appearance of this trait. By contrast, high smoking concordances are shared by both monozygotic and dizygotic twins, indicating that genetic differences do not affect the appearance of the trait, but rather, the trait is influenced by environmental factors common to twins brought up together.

(Reference: F. Conterio and B. Chiarelli, 1962, *Heredity*, **17**:347–359.)

10-17. One possible set of values for percentage concordances are the following:

Trait	Factors Determining Trait	Identical Twins Reared Together	Identical Twins Reared Apart	Fraternal Twins Reared Together	Fraternal Twins Reared Apart
(a)	only genetic	95	95	25	25
(b)	only environmental	75	25	75	25
(c)	1/2 genetic: 1/2 environmental	75	50	50	25

(a) Note that when the determining factors are *only genetic*, identical twins have the same concordances whether they are reared together or apart, and these concordances are considerably higher than those of fraternal twins.

(b) When the factors are *only environmental* there is no difference in concordances between identical and fraternal twins, but whether they are reared together or apart is the sole source of the difference in concordances.

(c) When the genetic and environmental effects on the appearance of the trait are *equal*, identical twins reared together (possessing both genetic and environmental similarities) have greater concordances than fraternal twins reared together (possessing fewer genetic similarities), and identical twins reared apart (possessing mainly genetic similarities) have greater concordances than fraternal twins reared apart (which possess fewer genetic similarities and none or few environmental similarities). As would be expected, the suggested concordances also show that twins reared together are more concordant than twins reared apart since environmental determinants are involved. Furthermore, the percentage concordance for identical twins reared apart (mainly genetic similarity) is equal to the percentage concordance for fraternal twins reared together (mainly environmental similarity), since there is a presumed equality between the genetic and environmental factors determining the trait.

11

Gene Interaction and Lethality

(References to "Examples" refer to text Table 11-1.)

11-1. The homozygous "recessive" parent in all the following cases is assumed to have the genotype *aabb* of box 16 in text Table 11-1. That is, for Examples 2 and 3, the *aabb* phenotypes would be horned white, and *BBNN*, respectively. Note that the backcross of F_1 (*AaBb*) × homozygous recessive (*aabb*) yields, in every case, four genotypes in equal ratio: 1 *AaBb*:1 *Aabb*:1 *aaBb*:1 *aabb*. The phenotypic appearance of these four genotypes will depend on the interactions that occur in each particular example.

Example (Table 11-1)	Phenotypes of Backcross Progeny
1	1 yellow round:1 yellow wrinkled:1 green round 1 green wrinkled
2	1 polled roan:1 polled white:1 horned roan: 1 horned white
3	1 ABMN:1 ABNN:1 BBMN:1 BBNN
4	1 walnut:1 rose:1 pea:1 single
5	1 agouti:1 black:2 albino
6	1 purple:3 white
7	2 white:1 yellow:1 green
8	3 triangular:1 ovoid
9	3 white:1 color
10	1 disc:2 sphere:1 long
11	1 partly rough:1 full rough:2 smooth
12	1 sooty:2 black:1 jet
13	1 sooty:2 black:1 dark sooty
14	1 glandular:3 glandless
15	1 shade 5:1 shade 3:1 shade 2:1 white (0)

11-2. A minimum of two successive crosses is necessary to obtain single-combed fowl: pea (*aaBB*) × rose (*AAbb*) → walnut *(AaBb)* × walnut (*AaBb*) → 1/16 single (*aabb*). Since the single-combed birds are homozygous recessive for both gene pairs, single × single should be pure-breeding.

11-3. (a) The absence of white-fruited plants in the F_1 and their relatively infrequent appearance in the F_2 indicates that it is a recessive trait. If it were caused by a single gene difference, it should appear in the F_2 with a frequency of 1/4. It is obviously more infrequent than that (1/32) and is probably caused by two gene pairs in which both must be homozygous recessive to produce white-fruited plants. (If it were caused by recessives at three gene pairs, the expected frequency of white-fruited plants would be 1/64.)

(b) Cross the F_1 to the white-fruited plants. A two-gene difference should produce a ratio of 3 blue:1 white ($AaBb \times aabb \rightarrow 1\ AaBb$:1 $Aabb$:1 $aaBb$: 1 $aabb$).

(Reference: I.V. Hall and L.E. Aalders, 1963, *Can. J. Genet. Cytol.,* **5**:371-373.)

11-4. (a) Judging from the 9/16:3/16:4/16 F_2 ratio, there are two gene pairs segregating.

(b) If we designate the alleles at the two gene pairs as A, a, and B, b, then the F_2 phenotypic ratios can be explained as resulting from interaction in which the homozygous recessive effect at one gene pair for short-awned (aa – –) is epistatic to the homozygous recessive effect at another gene pair for long-awned (– – bb). That is, the $aabb$ genotype is short-awned (see also Example 5, text Table 11-1).

$$
\begin{array}{lll}
9/16\ A\text{–}B\text{–} & = & \text{hooded} \\
3/16\ A\text{–}bb & = & \text{long-awned} \\
3/16\ aaB\text{–} & = & \text{short-awned} \\
1/16\ aabb & = & \text{short-awned}
\end{array} \Bigg\} = 4/16
$$

(c) Pure-breeding hooded stock $= AABB$. Pure-breeding short-awned stock $= aabb$. F_1 hooded $= AaBb$.

11-5. (a) Two gene factors seem to be involved here, producing a ratio of 12:3:1 (expected numbers based on a total of 560 plants is 420:105:35). As in Example 7, one effect (black in this case) is determined by dominance at one gene pair (e.g., A and a) and is epistatic to the effects at the other gene pair (gray and white, e.g., B and b). Thus, all F_1 individuals are heterozygous Aa and appear black, whereas only 3/4 of the F_2 carry the A allele and are black.

(b) Among the nonblack aa genotypes in the F_2, two kinds are possible, B- or bb, the former being gray and the latter white.

(Reference: H. Nilsson-Ehle, 1909, *Acta Univ. Lund Series 2,* **5**:21-122.)

11-6. The hypothesis of 9:6:1 leads to expected numbers of 9/16 $\times$ 37 = 21, 6/16 $\times$ 37 = 14, 1/16 $\times$ 37 = 2. Chi-square equals .24, which at 2 degrees of freedom, has a probability between .70 and .90. This indicates that the observed results fit in quite well with a 9:6:1 ratio.

11-7. The F_2 ratios indicate that cross *1* obviously involved segregation at one gene pair (e.g., *Pp*), whereas crosses *2* and *3* each involve segregation at two gene pairs. The fact that the F_1 of cross *2* is purple and the F_1 of cross *3* is colorless indicates that segregation at different gene pairs is involved in these crosses, and the colorless plant in cross *3* is probably homozygous for a color inhibitor (e.g., *II*). On the other hand, the colorless plant in cross *2* produces one fourth colorless plants in the F_2, indicating that colorless, in this case, is produced by homozygosity for a recessive gene (e.g., *cc*). On this basis, the crosses can be genotypically described as follows:

Cross *1*: P_1 : *PP* (purple) $\times$ *pp* (red)
F_1 : *Pp* (purple)
F_2 : 3/4 *P-* (purple): 1/4 *pp* (red)

Cross *2*: P_1 : *PPCC* (purple) $\times$ *ppcc* (colorless)
F_1 : *PpCc* (purple)
F_2 : 9/16 *P-C-* (purple):3/16 *ppC-* (red):3/16 *P-cc* (colorless):
1/16 *ppcc* (colorless)

Cross *3*: P_1 : *PPii* (purple) $\times$ *ppII* (colorless)
F_1 : *PpIi* (colorless)
F_2 : 12/16 *I-* (colorless):3/16 *P-ii* (purple):1/16 *ppii* (red)

(a) *PpCCii* $\times$ *PpCcii* $\rightarrow$ 3/4 purple:1/4 red
(b) *PpCCii* $\times$ *PpCCIi* $\rightarrow$ 3/8 purple:1/8 red:4/8 colorless
(c) *PpCcii* $\times$ *PpCCIi* $\rightarrow$ 3/8 purple:1/8 red:4/8 colorless

11-8. (a) The phenotypically AB daughter that appears in generation III must have received her *B* gene from her phenotypically O mother since she obviously obtained her *A* gene from her father. (Where else could the *B* gene have come from?) We can therefore assume that the mother (II-6) must be of the Bombay phenotype (*hh*). The twins (II-3 and II-4) may also have the Bombay phenotype, but the information given is not sufficient to be certain of this. (To decide whether they are *hh*, we would have to know the blood types of their offspring.) The mating in generation I therefore appears to be between two heterozygotes for the Bombay gene (*Hh* $\times$ *Hh*), and their blood group genotypes may have been *OO* and *BB*.

(b) The Bombay gene *cannot* be allelic to the *ABO* genes because the known Bombay phenotype in generation II (II-6) is homozygous for Bombay (*hh*) yet nevertheless passes on her *B* allele to one of her daughters. Thus, the *H* and *h* alleles are at a separate gene pair from *A, B,* and *O*.

(Reference: R.R. Race and R. Sanger, 1968, *Blood Groups in Man*, 5th ed. F.A. Davis, Philadelphia.)

11-9. (a) The cross *AaBb* $\times$ *AAbb* (or *AaBb* $\times$ *aaBB*) should produce 1/2 blue: 1/2 purple.

(b) 9/16 blue:6/16 purple:1/16 scarlet. (The Hagiwara data are given in H. Matsuura, 1929. *A Bibliographical Monograph on Plant Genetics.* Tokyo Imperial Univ. Press, Tokyo, pp. 156 and 158.)

11-10. (a) *AaBb* × *aabb*   (d) *AaBb* × *Aabb*
     (b) *Aabb* × *aaBb*   (e) *Aabb* × *aabb*
     (c) *AABb* × *AABb*

11-11. (a) Since red appears in an F_2 frequency 1/16, we can hypothesize that there are two gene pairs involved, and white is caused by the presence of a dominant allele at either of the two gene pairs (e.g., *A-bb, aaB-,* and *A-B-* = white).

    (b) According to this hypothesis, one strain of white plants is homozygous dominant for one of the two segregating gene pairs (e.g., *AAbb*), and the other white strain is homozygous dominant for the other segregating gene pair (e.g., *aaBB*). The F_1 is *AaBb* and consequently white, whereas the red F_2 plants are homozygous recessive for both gene pairs, *aabb* (see also Example 8).

(Reference: J. L. Brewbaker, 1962, *J. Hered.,* **53**:163-167.)

11-12. The chi-squares and associated probabilities given by Fuchs et al. are as follows:

		Numbers			
	Ratio	Glandular	Glandless	χ^2	Probability
	observed	89	36		
(a)	11:5 expected	85.9	34.1	0.35	.50–.60
(b)	15:1 expected	117.2	7.8	108.48	< .01
	13:3 expected	101.6	23.4	5.94	.01–.03
	9:7 expected	70.3	54.7	111.35	< .01

Note that the observed data fit the 11:5 expected ratios quite well, whereas the other F_2 ratios produce χ^2 all with probabilities of less than .05.

(Reference: J.A. Fuchs, J.D. Smith, and L.S. Bird, 1972, *J. Hered.,* **63**:300-303.)

11-13. (a) Since there are no genotypes other than yellow or green, among either the parents or the progeny, and the heterozygous F_1 is phenotypically like one of the parents, we can consider dominance as complete.

    (b) The 3:1 F_2 ratios produced in crosses I and II indicate that only single gene pair differences are involved in these crosses, although there is apparently a different gene pair segregating in each of the crosses. For example, we can designate the yellow parental strain (W) as *AABB* and the green parental strain (Z) as *aaBB*. Cross I is therefore *AABB* × *aaBB*, producing an F_1 *AaBB* that is yellow (*A* is dominant to *a*), and the F_2 segregates as expected, 3 *A–BB* (yellow):1 *aaBB* (green). Cross II, on the other hand, involves the *B* gene pair; that is, *aabb* (yellow strain Y) × *aaBB* (green strain Z) and produces an F_1 that is all green (*aaBb*). (Note that the yellow color pattern seems to depend on either the presence of an *A* allele at the *Aa* gene pair or the absence of a *B* allele at the *Bb* gene pair; *A*–is yellow, *aaB*– is green, and *aabb* is yellow.) According to our notation, the F_2 produced by cross II is in the expected ratio of 3 green (*aaB*–):1 yellow (*aabb*). The use of two different gene pairs to explain the results of crosses I and II helps also to explain the results of cross III in which different alleles at *both* pairs are segregating simultaneously: *AABB* (strain W) × *aabb* (strain Y) → *AaBb* × *AaBb* → 12 *A*– (yellow): 3 *aaB*– (green): 1 *aabb* (yellow). (See also Example 9, Table 11-1). Based on this notation, the answers to questions (c) through (h) are therefore as follows:

(c) *AABB*   (f) *aaBB*
(d) *AaBB*   (g) *aabb*
(e) *aabb*   (h) *aaBb* (2), *aaBB* (1)

11-14. (a) Flower color seems to be caused by a gene pair at which partial dominance appears, for example, *AA* = red, *Aa* = pink, *aa* = white. Flower shape is apparently caused by a gene at which normal is completely dominant over peloric, for example, *B*– = normal, *bb* = peloric. This is supported by the F_2 ratios of 1 red:2 pink:1 white, and 3 normal:1 peloric.

   (b) As in Example 2, the expected F_2 ratios are 6:3:3:2:1:1 or 88:44: 44:29:14:14. A chi-square test yields 1.26, which for 5 degrees of freedom means that the hypothesis is not rejected (.90 < probability < .95).

(Reference: E. Baur, 1910, *Z. Induk. Abst. U. Vererbung.*, **3**:34-98.)

11-15. 13/16 (See Example 9.)

11-16. This pedigree evidently involves the segregation of two different gene pairs since the gene pair segregating in the left-hand family, although similar in recessive effect to the gene pair segregating in the right-hand family, nevertheless produces normal offspring in the mating between the two types of homozygous recessive. One way of explaining this is to assume that the trait can be caused by homozygosity at one gene pair (e.g., *aa*) or at another (e.g., *bb*) but the presence of a dominant gene at both of these gene pairs (e.g., *A–B–*) is sufficient to produce the normal phenotype (see complementary genes, Example 6, text p. 184). Thus the left-hand family can be considered as segregating for *A* and *a*, and the right hand family as segregating for *B* and *b*; deaf individuals are *aa* – – or – –*bb*. On this basis, the genotypes of the numbered individuals would be (1) *AaBB* (2) *AABb* (3) *aaBB* (4) *AAbb* (5) *AaBb*.

11-17. Apparently single gene differences are segregating in each of the first two crosses, and alleles at two gene pairs are segregating in cross c. In cross a, spiny-tip is obviously dominant (e.g., S) over spiny (e.g., s), but in crosses b and c piping is dominant (e.g., P) over both spiny and spiny-tip (e.g., p). Thus, spiny-tip can be described genotypically as $ppSS$ and spiny as $ppss$. The fact that no spiny F_2 appear in cross b indicates that piping must be $PPSS$. Thus,

      (a) $ppSS \times ppss$.
      (b) $PPSS \times ppSS$.
      (c) $PPSS \times ppss$ (see Example 7).

(Reference: J.L. Collins and K.R. Kerns, 1946, *J. Hered.*, 37:123–128.)

11-18. (a) $PpSs \times ppss \rightarrow$ 1/2 piping:1/4 spiny-tip:1/4 spiny
      (b) $PpSs \times ppSs \rightarrow$ 1/2 piping:3/8 spiny-tip:1/8 spiny

11-19. (a) If we designate the symbols for *black* and *ebony* as b and e, respectively (wild-type alleles are b^+ and e^+), then the cross between the two stocks is $bb\ e^+e^+ \times b^+b^+\ ee \rightarrow b^+b\ e^+e$, all wild type.
      (b) $F_1 \times F_1 = b^+b\ e^+e \times b^+b\ e^+e$

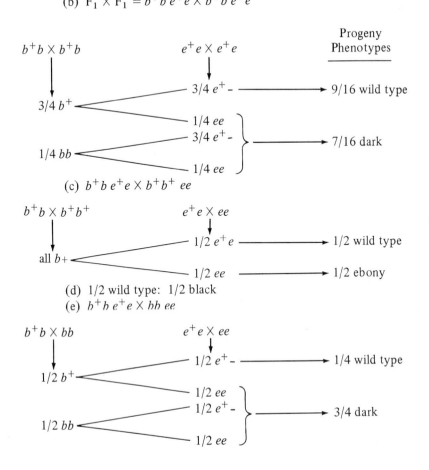

      (c) $b^+b\ e^+e \times b^+b^+\ ee$

      (d) 1/2 wild type: 1/2 black
      (e) $b^+b\ e^+e \times bb\ ee$

11-20. There appear to be two pairs of genes involved in color determination, one of which determines whether any color appears at all (white vs. yellow-green) and the second of which determines the kind of color (yellow vs. green). The fact that white plants may occasionally produce colored plants, but never the reverse, indicates that the absence of color is caused by a dominant "epistatic" allele (e.g., A) and that white plants may be heterozygous for this allele (Aa) but colored plants must be homozygous recessive (aa). Thus, white heterozygotes will occasionally produce colored offspring from a mating $Aa \times Aa$. In cross II such a mating produces offspring in the expected ratio of 3 white (A-):1 colored (aa). On the other hand, for those individuals that are colored (aa), yellow appears to be dominant over green, as seen in the F_2 data for cross I. We can therefore designate the second gene pair as yellow $= B$ and green $= b$, and the parents in cross I are obviously $aaBB$ (yellow) $\times$ $aabb$ (green), whereas the cross II parents are $AAbb$ (white) $\times$ $aaBB$ (yellow). (These data are essentially those used to derive Example 7 in Table 11-1.) In cross II, the expected F_2 numbers of white, yellow, and green on a 12:3:1 basis are 154:38:13. A statistical test yields a χ^2 of .69, which is certainly low enough to permit the hypothesis of a 12:3:1 ratio to be accepted.

(Reference: E.W. Sinnott and G.B. Durham, 1922, *J. Hered.*, **13**: 177-186.)

11-21. (a) The data in this problem correspond to those of Example 6. Dwarf is produced by homozygous recessive genes at either one or both of two gene pairs, for example, strain *1* is $aaBB$, and strain *2* is $AAbb$. Each strain when crossed to wild type ($AABB$) will segregate only for one pair of genes and therefore produce an F_2 with only 1/4 dwarf. In the cross strain *1* $\times$ strain *2*, both pairs of genes are segregating, and the F_2 is therefore 7/16 dwarf.

    (b) $AaBb \times aaBB \rightarrow$ 1 normal:1 dwarf

    (c) $AaBb \times AAbb \rightarrow$ 1 normal:1 dwarf

    (d) $AaBb \times AaBB \rightarrow$ 3 normal:1 dwarf

11-22. P (pink) is dominant over p (yellow); D (dark) is dominant over d (transparent). The fact that the cream-colored snail with transparent bands produced all pink offspring when mated to a recessive yellow individual indicates that the cream-colored parent was homozygous pink, that is, of genotype $PPdd$. In other words, there is an epistatic interaction caused by dd that produces cream color in such genotypes. Because all of the F_1 are also dark-banded, the yellow dark-banded parent of the cross must have been $ppDD$; that is, the cross was $PPdd \times ppDD \rightarrow PpDd$ (pink, dark-banded). The $F_1 \times F_1$ cross mentioned in the problem was $PpDd \times PpDd$ and produced only one F_2 individual $P\text{-}dd$ (cream with transparent bands).

(Reference: J. Murray, 1963, *Genetics,* **48**:605-615.)

11-23. If you crossed red $\times$ white animals, the single-gene hypothesis would predict only roan offspring since RR (red) $\times$ rr (white) $\rightarrow Rr$ (roan). On the other hand, the two-gene hypothesis would predict that roan offspring would be exclusively produced *only* if the parents were $RRpp$ (red) $\times$ $rrPP$ (white) $\rightarrow$

RrPp (roan). Should two gene pairs be involved, and the red and white parents possess any of the four other possible genotypic combinations, ① *RRpp* × *rrpp* or ② *RRpp* × *rrPp* or ③ *Rrpp* × *rrpp* or ④ *Rrpp* × *rrPp*, then the offspring would be all red ① , or red and roan ② , or red and white ③ , or red, roan, and white ④ . That is, if two gene pairs are involved, large populations of cattle would not be expected to possess only one genotype for red (*RRpp*) and white (*rrPP*), but some of the other possible combinations should also occur. In other words, the presence of offspring other than roan from a large number of red × white crosses would be expected if two gene pairs are interacting to produce these coat colors. By contrast, the single-gene hypothesis states that the red phenotype can only be *RR* and the white phenotype only *rr*, and *only roan* offspring would be expected from such crosses. Wright collected data from various herd books and showed that, except for some misclassifications, only roan offspring were produced from red × white matings, thus supporting the single gene hypothesis.

Note also that a cross of roan × roan according to the single-gene hypothesis is, by definition, a cross between heterozygotes, *Rr* × *Rr*, and should produce offspring in the ratio 1/4 red (*RR*):2/4 roan (*Rr*):1/4 white (*rr*). The two-gene pair hypothesis suggests that such crosses will often consist of at least some homozygotes, for example, *RRPP* × *RRPP* or *RRPP* × *RrPP* or *RRPP* × *RRPp* or *RRPP* × *RrPp*. In each of these cases, the offspring are *all* roan, although other crosses (e.g., *RrPp* × *RrPp*) would produce some red and white offspring as well. Thus, the two-gene pair hypothesis predicts a large preponderance of roan offspring from random roan × roan matings, and the single-gene pair hypothesis predicts a fairly exact ratio of 1/4 red:2/4 roan:1/4 white. The data collected by Wright for roan × roan crosses also supported the single-gene pair hypothesis.

(Reference: S. Wright, 1917, *J. Hered.* **8**: 521–527.)

11–24. The phenotype of the F_1 indicates that the allele or alleles that determine yellow are dominant over those that determine green. The F_2 results accord with the hypothesis that there are a number of different independently assorting gene pairs that can produce a green phenotype when homozygous for the recessive allele at any one of these gene pairs. The frequencies of yellow and green phenotypes can then be calculated as follows:

A cross in which only one gene pair is segregating (e.g., $Aa \times Aa$) and all other gene pairs affecting color are homozygous for the dominant *yellow* alleles (e.g., *AaBBCC* × *AaBBCC*) will produce yellow offspring in the usual ratio of $(3/4)^1 = 3/4$, and green in the ratio of $1 - (3/4)^1 = 1/4$. If two gene pairs are segregating in the cross (e.g., *AaBbCC* × *AaBbCC*), then $(3/4)^2 = 9/16$ are yellow and $1 - (3/4)^2 = 7/16$ are green. For three gene pairs (e.g., *AaBbCb* × *AaBbCc*) $(3/4)^3 = 27/64$ are yellow, and $1 - (3/4)^3 = 37/64$ are green. The answers are therefore as follows:

(a) 3   (b) 1   (c) 2

As the number of gene pairs segregating for color increases in this example, the chances for a green recessive phenotype to appear among the F_2 increases, since homozygosity for the recessive allele at *any* of the increasing number of segregating genes will cause appearance of the recessive phenotype.

11-25. If we designate strain X as *aabbcc*, strain Y as *aaBBCC*, and strain Z as *aabbCC*, we can see that strain X shares a pair of homozygous recessive genes in common with strains Y and Z. Thus, green color caused by homozygous recessives at any of these three gene pairs will appear in all F_1 and F_2 progeny of crosses between strain X and strain Y or Z. The answers to questions (a) to (d) are therefore all green progeny. If different genes are assumed to be segregating in each strain, that is, strain X is designated *aabbccDDEE*, strain Y as *AABBCCddEE*, and strain Z as *AABBCCddee*, four and five pairs of genes are segregating, respectively, in the crosses X X Y and X X Z, and the answers are
  (a) All yellow (*AaBbCcDd*)   (c) All yellow (*AaBbCcDdEe*)
  (b) $1 - (3/4)^4 = 175/256$ (d) $(3/4)^5 = 243/1024$

11-26. The parental cross is *AACCRRpp* X *AACCrrPP*, yielding an F_1 which is *AACCRrPp*. Essentially the F_1 X F_1 cross can therefore be represented as segregating for differences at only two gene pairs, *RrPp* X *RrPp*, or

11-27. We can diagram the answer as follows:

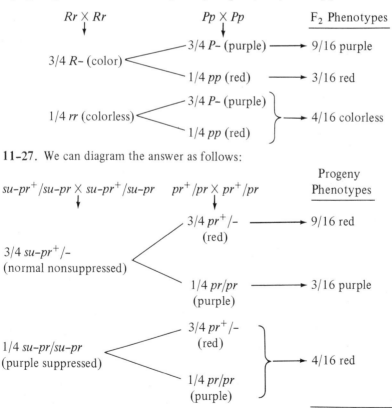

total: 13/16 red: 3/16 purple

11-28. The expected phenotypic ratio for the progeny of a cross between F_1 individuals differing at a single pair is 3 dominants:1 recessive. The observed F_2 ratio in this case, 2 platinum:1 silver, is similar to the 2:1 ratio observed for recessive lethals (see text p. 188); that is, the platinum homozygote is lethal. The observed ratio is therefore that of platinum heterozygotes (2) to silver homozygotes (1).

(Reference: I. Johansson, 1947, *Hereditas*, **33**:152-174.)

11-29. The observed number of defective eggs is exactly 1/16 of the total. Thus, segregation at two gene pairs seems to be involved, and we can assume that the double recessive homozygote *aabb* is lethal.

(Reference: V.S. Asmundsen, 1942, *J. Hered.*, **33**:328-330.)

11-30. Yellow and black segregate as alleles of agouti, with the yellow effect dominant over agouti (homozygous yellow is apparently lethal, see also text Fig. 11-5) and agouti dominant over black. Using the gene symbols on text p. 186, we can designate yellow as A^Y, agouti as A, and black as a. On this basis, the crosses can be written as follows:

				F_1 Progeny
Cross I:	$A^Y A$	$\times$	$A^Y A$	$1\,A^Y A^Y$ (lethal): $2A^Y A$:$1AA$
Cross II:	$A^Y A$	$\times$	AA	$1\,A^Y A$: $1\,AA$
Cross III:	$A^Y a$	$\times$	aa	$1\,A^Y a$: $1\,aa$
Cross IV:	aa	$\times$	AA	all Aa

The answers are therefore as follows:
(a) $A^Y A \times A^Y A \rightarrow$ 2/3 yellow:1/3 agouti ($A^Y A^Y$ are lethal.)
(b) $Aa \times A^Y A \rightarrow$ 1/2 yellow:1/2 agouti
(c) $Aa \times A^Y a \rightarrow$ 1/2 yellow:1/4 agouti:1/4 black

11-31. (a) Assuming that there is a single gene difference involved and that green (A) is dominant to virescent (a) a backcross of the type $Aa \times aa$ would be expected to produce 1/2 virescent. The shortage of virescent may then be ascribed to lethality.

(b) A cross of the $F_1 \times F_1$ ($Aa \times Aa$) should produce 3 green:1 virescent. A shortage of virescent would again testify to lethality. As a further test, one could score the germination abilities of seeds of normal tomato plants and of those that are segregating for virescent homozygotes.

11-32. (a) If we use the symbols A^Y and T^B for the *yellow* and *brachyury* alleles, respectively, and A^+ and T^+ for their respective wild-type alleles, we can symbolize the cross as $A^Y A^+ \, T^B T^+ \times A^Y A^+ \, T^B T^+$. Ratios of expected phenotypes among the progeny can then be derived by the following method:

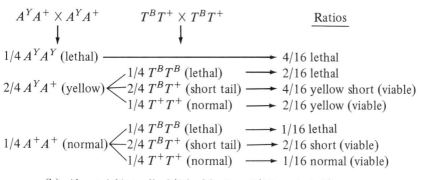

$A^Y A^+ \times A^Y A^+$ $T^B T^+ \times T^B T^+$ Ratios

$\downarrow$ $\downarrow$

$1/4\ A^Y A^Y$ (lethal) $\longrightarrow$ 4/16 lethal

$\qquad\qquad\qquad$ $1/4\ T^B T^B$ (lethal) $\longrightarrow$ 2/16 lethal

$2/4\ A^Y A^+$ (yellow) $\longleftarrow$ $2/4\ T^B T^+$ (short tail) $\longrightarrow$ 4/16 yellow short (viable)

$\qquad\qquad\qquad$ $1/4\ T^+ T^+$ (normal) $\longrightarrow$ 2/16 yellow (viable)

$\qquad\qquad\qquad$ $1/4\ T^B T^B$ (lethal) $\longrightarrow$ 1/16 lethal

$1/4\ A^+ A^+$ (normal) $\longleftarrow$ $2/4\ T^B T^+$ (short tail) $\longrightarrow$ 2/16 short (viable)

$\qquad\qquad\qquad$ $1/4\ T^+ T^+$ (normal) $\longrightarrow$ 1/16 normal (viable)

(b) About 4 (Actually 9/16 of 8, since 7/16 are lethal.)

11-33. The simplest explanation is to assume that the original stock is hetero-zygous for genes A and B, each located on a different member of a pair of homo-logous chromosomes, so that when they are mated to wild type, each offspring gets only one of the two homologues, A or B, but never both. The fact that the original stock is always heterozygous for these two genes indicates that they are maintained in this condition by a balanced lethal system (see text pp. 188–190, and also p. 437).

11-34. (a) *Hhss* $\times$ *hhSs*. The ratio will be 3 normal:1 hairless (*Hhss*), but the normals will consist of three different genotypes, *hhss*, *hhSs*, and *HhSs*.

$\quad$ (b) *hhss* $\times$ *Hhss* $\rightarrow$ $\quad$ 1 normal:1 hairless
$\qquad$ *hhSs* $\times$ *Hhss* $\rightarrow$ $\quad$ 3 normal:1 hairless
$\qquad$ *HhSs* $\times$ *Hhss* $\rightarrow$ $\quad$ 2 normal:1 hairless
$\qquad$ *Hhss* $\times$ *Hhss* $\rightarrow$ $\quad$ 1 normal:2 hairless

$\quad$ (c) *hhss* $\times$ *hhSs* $\rightarrow$ $\quad$ all normal
$\qquad$ *hhSs* $\times$ *hhSs* $\rightarrow$ $\quad$ all normal
$\qquad$ *HhSs* $\times$ *hhSs* $\rightarrow$ $\quad$ 5 normal:1 hairless
$\qquad$ *Hhss* $\times$ *hhSs* $\rightarrow$ $\quad$ 3 normal:1 hairless

$\quad$ (d) *hhss* $\times$ *hhss* $\rightarrow$ $\quad$ all normal
$\qquad$ *hhSs* $\times$ *hhss* $\rightarrow$ $\quad$ all normal
$\qquad$ *Hhss* $\times$ *hhss* $\rightarrow$ $\quad$ 3 normal:1 hairless
$\qquad$ *Hhss* $\times$ *hhss* $\rightarrow$ $\quad$ 1 normal:1 hairless

$\quad$ (e) Since there are four F_1 genotypes, there are 16 possible mating com-binations. Twelve of these have been considered above. The remaining four are:
$\qquad$ *hhss* $\times$ *HhSs* $\rightarrow$ $\quad$ 3 normal:1 hairless
$\qquad$ *hhSs* $\times$ *HhSs* $\rightarrow$ $\quad$ 5 normal:1 hairless
$\qquad$ *HhSs* $\times$ *HhSs* $\rightarrow$ $\quad$ 7 normal:2 hairless
$\qquad$ *Hhss* $\times$ *HhSs* $\rightarrow$ $\quad$ 2 normal:1 hairless

11-35. A general reduction in the frequency of hairless progeny would be expected in those instances in which the males are heterozygous for the suppressor gene.

12

Sex Determination and Sex Linkage in Diploids

12-1. Since the F_1 were all broad-leaved, the maternal broad-leaf effect is dominant over the paternally transmitted narrow leaf. If females in this species were heterogametic (e.g., XY) and males were homogametic (e.g., XX), we would expect the hemizygous XY F_1 females to inherit their X chromosome from their presumed homogametic XX father and, therefore, to be narrow-leaved. Since the F_1 females are broad-leaved, they must inherit an X chromosome from their mother as well, that is, the F_1 females are probably XX but heterozygous for the broad-leaf and narrow-leaf alleles, and the males are XY. This view of homo-gametic females and heterogametic males is supported by the F_2 results in which the males show that they are of two types, that is, the products of heterozygous XX females. The F_2 females are all broad-leaved, since they inherit one of their two X chromosomes from their XY F_1 father who carries only a broad-leaf allele on his X chromosome.

(Reference: G.H. Shull, 1914, *Z. Indukt. Abst. und Vererb.*, **12**:265-302.)

12-2. Since *B. alba* is monoecious, it has by definition no genetic differences between the sexes, and we can therefore assume it is homogametic. In *B. dioica*, the male is obviously the heterogametic sex since it is the male pollen that produces the two different sexual forms when fertilizing the homogametic *B. alba*. On this basis, we can describe the *B. dioica* male as heterogametic XY, and the *B. dioica* female as homogametic XX, whereas the monoecious *B. alba* is homogametic XX. Thus, XX × XX fertilizations produce only females, whereas XX × XY fertilizations produce both sexes.

(Reference: C. Correns, 1907. *Die Bestimmung und Vererbung des Geschlechts nach neuen Versuchen mit höheren Pflanzen.* Borntraeger, Berlin.)

12-3. The cross occurs between chromosomal XY male tissues, and the YY progeny are also males:

$$\text{XY} \times \text{XY} \rightarrow \quad 1 \text{ XX:} \quad 2 \text{ XY:} \quad 1 \text{ YY}$$
$$\qquad\qquad\qquad \text{(females)} \quad \text{(males)} \quad \text{(males)}$$

(Reference: H.C. McPhee, 1925. *J. Agric. Res.*, **31**:935-943.)

12-4. The answer to this is believed to lie in the degree of competition between male and female pollen. In *Lychnis*, the male is heterogametic, and excess pollination causes pollen competition and the relative success of female pollen. Sparse pollination reduces pollen competition and enables both types of pollen to have equal chances for fertilization. In *Fragaria* there is, of course, no competition between male and female pollen since the male is homogametic.

(Reference: C. Correns, 1917. *Sitzungsber. der Königl. Preuss. Akad. der Wissensch.*: 685-717.)

12-5. The fact that both males and females result from a cross between two "chromosomal" females indicates that the females in this species are *not* the homogametic sex, otherwise only females would be produced. Thus, the females are heterogametic (often called ZW), and the cross can be symbolized as

$$ZW \times ZW \rightarrow \quad 1\ ZZ: \quad 2\ ZW: \quad 1\ WW$$
$$\text{(males)} \quad \text{(females)} \quad \text{("females")}$$

This accords with the observed ratios of 1♂:3♀.

(Reference: R.R. Humphrey, 1945, *Am. J. Anat.*, 76:33-66.)

12-6. The male sex is homogametic and the female sex is heterogametic. The reason for this is that the estradiol-treated "females" that produce only male offspring *must* represent sex-reversed males. If these sex-reversed males were heterogametic (e.g., XY), then the mating XY"♀" × XY♂ would certainly *not* be expected to produce exclusively males (XY or YY) but should also produce females (XX). On the other hand, if the sex-reversed males are homogametic (XX), then the mating of XX"♀" × XX♂ would produce only XX offspring who are phenotypic males (unless treated with estradiol). It is this latter that we observe. Thus, half of the estradiol-treated larvae become sex-reversed (but homogametic XX) individuals that act as "females" but only produce male (XX) offspring; and the other half of the estradiol-treated larvae remain as normal heterogametic XY females who are not affected by the treatment, and who produce both male (XX) and female (XY) offspring.

(Reference: C.Y. Chang and E. Witschi, 1956, *Proc. Soc. Exp. Biol. Med.*, 93:140-144.)

12-7. In the female, 18 autosomes (9 pairs = 18A) would produce meiotic gametes of 9A, and 24X chromosomes (12 pairs = 24X) would produce meiotic gametes of 12X, that is, the gametic chromosome constitution is 9A + 12X. In the male (18A + 12X + 6Y), two kinds of gametes are produced, 9A + 12X and 9A + 6Y, based on an even division of autosomes and the segregation of X's and Y's to opposite poles. Obviously the egg-sperm combination of (9A + 12X) + (9A + 12X) produces females, and the combination (9A + 12X) + (9A + 6Y) produces males.

12-8. Parthenogenesis occurs when unfertilized eggs can develop into normal adults (see text p. 201) and, in diploids, is usually associated with a doubling of the haploid number of chromosomes in the oocyte.

(a) In *Drosophila*, the female is homogametic; thus, parthenogenesis, by chromosome doubling among oocytes containing the X chromosome, leads to XX individuals that are female. In birds, on the other hand, it is the male that is homogametic (e.g., XX) and the female that is heterogametic (e.g., XY). Parthenogenesis by chromosome doubling among XY bird oocytes therefore leads to XX individuals that are male or to YY zygotes that are missing a necessary X and are therefore inviable.

(b) Nondisjunction of the autosomes but not of the X chromosome can produce an egg with two sets of autosomes and only one X. If such an egg developed parthenogenetically, it would be an XO male (see text p. 210).

12-9. A stock that is homozygous for the *faded* gene would produce hemizygous females having a phenotype (faded) different from homozygous males (white).

(Reference: W.F. Hollander, 1942, *J. Hered.*, 33:135–140.)

12-10. If one third of the eggs are unfertilized, these are haploid, or males. The remaining two thirds will have been formed by the combination Xa/Xb (mother) $\times$ Xa (son), since the son derives its single X allele from its mother. The diploid progeny of this cross will therefore consist of half females (Xb/Xa) and half males (Xa/Xa), and the total sex ratio (including haploid males) will be $1/3\,\male + 2/3[1/2\male{:}1/2\female] = 2/3\male{:}1/3\female$.

12-11. (a) A 1:1 sex ratio is expected because, according to text Fig. 12–17, the females produced by the female-producing mothers are male- and female-producers in an approximate 1:1 ratio.

(b) There would be no expected difference in males but there would be an expected difference in females, since elimination of the X' maternal chromosome would lead to the production of only male-producers. Numerous findings suggest that it is the male X chromosome that is eliminated.

(Reference: C.W. Metz, 1938, *Am. Nat.*, 72:485–520.)

12-12. (a) 1 female:3 males. Of the phenotypic males, one third will actually be transformed females (XX *tra/tra*) and therefore sterile.

(b) 3 females:5 males. Of the phenotypic males, one fifth will actually be transformed and sterile XX individuals.

12-13. The female is genetically *sk sk Ts_3ts_3* and therefore has sterile ovaries in the ears but fertile ovaries in the tassel. The male is genetically *sk sk ts_3ts_3* and therefore has sterile ovaries in the ears but pollen-forming organs in the tassel. Sex is thus determined by the *Tassel seed$_3$* gene, and crosses are of the type $Ts_3ts_3 \female \times ts_3ts_3 \male$.

(References: D.F. Jones, 1934, *Genetics*, **19**:552-567; and R.A. Emerson, 1933, *Proc. Sixth Intern'l. Congr Genetics*, **1**:141-152.)

12-14. The female is of genotype *sk sk ts*$_2$ *ts*$_2$ with fertile ovaries in both ears and tassel. The male is genetically *sk sk Ts*$_2$ *ts*$_2$ with sterile ovaries in the ears but pollen-forming organs in the tassel. Crosses are thus consistently of the type *ts*$_2$ *ts*$_2$ ♀ × *Ts*$_2$ *ts*$_2$ ♂.

12-15. Female 1 is obviously homozygous for a sex-linked recessive mutation causing abnormal eyes $X^a X^a$ because all of her hemizygous sons ($X^a Y$) show the trait, whereas her heterozygous daughters ($X^a X^+$) have inherited the X^+ chromosome from their father and are therefore normal. Female 2 is heterozygous for a dominant gene producing abnormal eyes since half of her daughters show the trait. From the information given, there is no way of knowing whether the dominant gene is sex-linked or autosomal. (As a further test, the F_1 abnormal-eyed males should be mated to normal-eyed females. What results would you expect for the two alternatives?)

12-16. The three phenotypes appear to be caused by codominant alleles of a single gene that segregate in fairly predictable fashion, that is, *aa* = *a* phenotype; *bb* = *b* phenotype; *ab* = *ab* phenotype. Thus, the cross *ab* × *ab* produces all three phenotypes. Since males are hemizygous for the X chromosome, they would have to be either *a* or *b* but not *ab* if this were a sex-linked trait. The presence of *ab* males makes it obvious that this trait is caused by an autosomal gene. (Other variants of this enzyme, however, are known to be caused by sex-linked genes.)

(Reference: C.R. Shaw and E. Barto, 1965, *Science*, **148**:1099-1100.)

12-17. Sex-linked recessive. Note that this pedigree has the characteristics mentioned on text p. 207: (1) Females are the carriers of the disease. (2) The disease appears only among the sons of such carriers. (3) Males produced in the pedigree who are free of the disease do not pass it on.

12-18. It is probably caused by a sex-linked recessive gene because none of the daughters in the pedigree shows the trait, but it appears in exactly half of the sons of female I-2, who is most probably a heterozygote. One form of muscular dystrophy, Duchenne's, is of this type.

12-19. The cause is apparently a dominant sex-linked gene since all eight daughters of affected males show the disease but none of the eight sons of such males are affected.

(Reference: R.W. Winters, J.B. Graham, T.F. Williams, V.W. McFalls, and C.H. Burnett, 1958, *Medicine*, **37**:97-142.)

12-20. It appears to be caused by an autosomal dominant gene that is expressed only in males since, as shown on the right-hand side, it is passed on from son to son but never appears in females of the pedigree. If it were a sex-linked recessive, female III–6 should be carrying it on the X chromosome she inherited from her father. Yet, of her six sons, none shows the trait.

12-21. (a) No, since male offspring of affected males show the trait. (Males do not inherit their father's X chromosome.)

(b) No. For the same reason given in (a). Also, *all* female offspring of affected males would be expected to show the trait if it were an X-linked dominant since they inherit their father's X chromosome.

(c) No. Since females do not carry a Y chromosome, none of the daughters of an affected male would be expected to show the trait. This pedigree, however, shows one such affected daughter (III-12).

(d) Yes.

(e) No. Since the trait is assumed to be very rare, all individuals outside the pedigree (e.g., II–1) would not be expected to carry the gene, and their progeny should be normal.

12-22. (a) No. Since the female offspring of affected males show the trait, this would mean that all the normal females who mate with such males are heterozygous carriers of the sex-linked recessive gene. (In such cases, half of their sons would also be expected to show the trait.) However, the trait is rare, and such frequent heterozygous carriers are not expected. Also, if the trait is caused by a sex-linked recessive gene, *all* the male offspring of affected females should show the trait, but this obviously does not happen (III-8, III-11, III-13).

(b) Yes. All the female offspring of an affected male show the trait (daughters inherit their father's X) but *none* of the sons of such a male show the trait (sons do not inherit their father's X). Also, as expected, about one half of the sons and one half of the daughters of an affected female show the trait.

(c) No. Since females have no Y chromosome, none of the daughters of an affected male would be expected to show the trait, but there are many such affected females in this pedigree.

(d) This is very unlikely. If the trait were autosomal dominant, affected males should produce one half affected females and one half affected males. However, in this case *only* the daughters of affected males show the trait, but *none* of the sons.

(e) No. Since the trait is assumed to be very rare, all outside individuals marrying into the pedigree would not be expected to be carrying the trait, and the progeny of practically all of the matings given should be normal.

12-23. (a) No, since male offspring of affected males show the trait. (Males *do not* inherit their father's X chromosome.) Furthermore if the gene were a sex-linked recessive, *all* male offspring of affected females (I-1, II-4) would be expected to show the trait. But this is obviously not so. Also, the only way a female would show a sex-linked recessive trait would be if both parents transmitted it; that is, if her father were a carrier and showed the trait. Note, however, that the fathers of affected females II-4, III-2, and IV-3 do not show the trait.

(b) No. This is excluded for the same reasons as above. Also, *all* female offspring of affected males would be expected to show the trait since they always inherit their father's X chromosome. Female IV-5, however, is free of the trait.

(c) No, since none of the daughters would be expected to show the trait.

(d) Yes. Note that the trait appears only in offspring who have an affected parent. (It does not "skip" generations.)

(e) No. Since the trait is assumed to be very rare, all individuals who marry into the family (II-5, III-1, III-6) would *not* be expected to be carrying the gene, and we would therefore expect their progeny to be normal if the trait were caused by a homozygous recessive.

12-24. Colorblindness is obviously a fully recessive trait, appearing only in those hemizygous males carrying it on their X chromosome. G-6-pd deficiency is a partially dominant sex-linked trait, with females showing the three possible phenotypes (since they have three possible genotypes, gg, g^+g, g^+g^+) whereas males show only two possible phenotypes corresponding to their two genotypes (g or g^+). Xg blood type is apparently dominant ($a^+ > a^-$) since the daughters of II-1 show the trait (inheriting their father's X chromosome) but his sons do not show it (inheriting their father's Y chromosome). The genotypes of the individuals mentioned in the problem are as follows: I-1 $c\,g^+\,a^-$; I-2 $c^+\,c^+\,gg$ a^-a^-; II-1 $c^+g^+a^+$; II-2 $cc^+gg^+a^+a^-$; III-1 $c^+g\,a^-$; III-5 $cc^+gg^+a^+a^-$ or $c^+c^+gg^+a^+a^-$. (Pedigrees are from R. Sanger, 1965, *Can. J. Genet. Cytol.*, 7:179–188.)

12-25. (a) In moths, the female transmits its X chromosome to its sons, but not to its daughters. Maternal transmission of a sex-linked gene to daughters must therefore arise through loss of the paternal X chromosome early in the embryogenesis of a putative XX male by nondisjunction, thereby producing an XO female. XO females can also arise through loss of a paternal X chromosome in sperm formation (because of nondisjunction), which then fertilizes an X-bearing egg.

(b) If the two X chromosomes in the males of this stock are "attached" (see text pp. 210–211), only two types of sperm are produced, XX and O. The daughters are then consistently formed by an O-bearing sperm meeting an X-bearing egg. That is, X-chromosome transmission is from mother to daughter. (Would such a stock be self-perpetuating if triple X and OO zygotes are inviable or sterile?)

12-26. Nondisjunctional eggs in human, mouse, and *Drosophila* are XX and O. Fertilization of these eggs by normal sperm produces four exceptional genotypes, XXX, XXY, XO, YO, whose phenotypes will depend on the developmental effects of the sex chromosomes and their relationship to the autosomes (see text p. 210 and Chapter 21). Nondisjunctional eggs in XY heterogametic females are XY and O. Fertilization of such eggs with X-bearing sperm from homogametic XX males will produce XXY and XO zygotes whose phenotypes, again, are dependent on sex chromosome-autosome relationships.

12-27. The expected frequency of XXY females is 1 per 5000 since the XX egg (frequency 1/2500) is fertilized only half the time with a Y-bearing sperm. The frequency of XO males would be expected as 1 per 1200 since the O egg (frequency 1/600) is fertilized only half the time with X-bearing sperm.

12-28. (a) All males and females will be normal.
(b) Half of the sons will be hemophilic. All the daughters will be normal.

12-29. Males: c^+ = normal; c = colorblind
Females: $c^+ c^+$ = normal; $c^+ c$ = normal; cc = colorblind

12-30. (a) No. (b) No. (c) Yes. (d) No.

12-31. (a) Yes. (b) Yes. (c) No. (d) Yes.

12-32. (a) No. (b) Yes. (c) Yes. (d) Yes.

12-33. (a) Yes. (b) Yes. (c) Yes.

12-34. (a) c = colorblind (b) $c^+ c$ = normal

12-35. (a) 1/2 (b) 1/2

12-36. 1/2

12-37. The genotype of the mother is AO in respect to ABO blood type (she has the A phenotype yet produces O offspring) and $c^+ c$ in respect to color-blindness (she herself has normal color vision, c^+, but must also carry c since her two sons are colorblind). According to the phenotypic descriptions given, possible male parent 1 is genotypically $AB\,c$, and possible male parent 2 must be genotypically $AO\,c^+$ (to account for his A phenotype and the presence of O offspring). The possible matings and offspring can be described as follows:

Matings		Offspring		
		Phenotype	Genotype	Male Parent
mother $AO\,c^+c$	(a)	A c♂	A- c	1 or 2
	(b)	O c♂	$OO\,c$	2
	(c)	A c♀	$AA\,cc$	1
male 1 male 2	(d)	B c$^+$♀	$BO\,c^+c$	1
$AB\,c$ $AO\,c^+$	(e)	A c$^+$♀	A- c^+-	1 or 2

12-38. (a) The parental cross can be represented as:

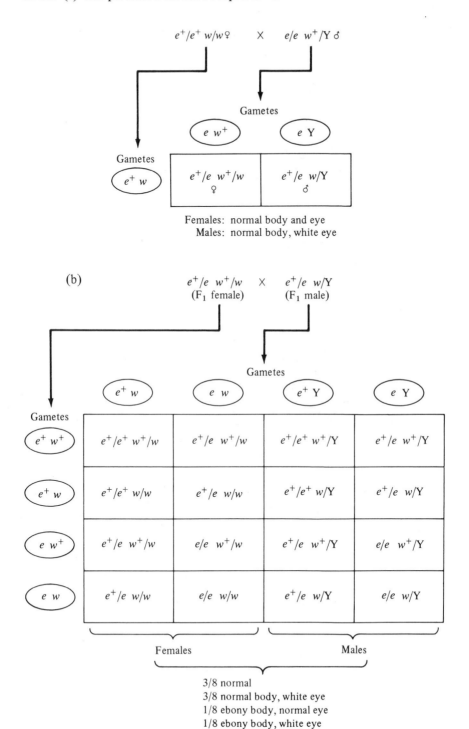

Females: normal body and eye
Males: normal body, white eye

3/8 normal
3/8 normal body, white eye
1/8 ebony body, normal eye
1/8 ebony body, white eye

12-39. The expected numbers are determined from the ratios given in the answer to part (b) of Problem 12-38, multiplied by the totals for males (551) and females (569).

      (a) Males: 206 wild type:206 white:69 ebony:69 white ebony

            Females: 213 wild type:213 white:71 ebony:71 white ebony

      (b) Chi-square for males is 1.67 and for females 2.06. In both cases, this shows that there is no significant deviation from the expected ratio.

12-40. (a) The parental genotypes can be represented as $G/G\ w^+/w^+$ ♀ × G^+/G^+ w/Y ♂ producing $G^+/G\ w^+/w$ females and $G^+/G\ w^+/Y$ males; that is, all of the F_1 show the grape phenotype because its effect is dominant in heterozygotes (G^+/G) and because none of the F_1 females (w^+/w) or males (w^+/Y) is white-eyed.

(b)

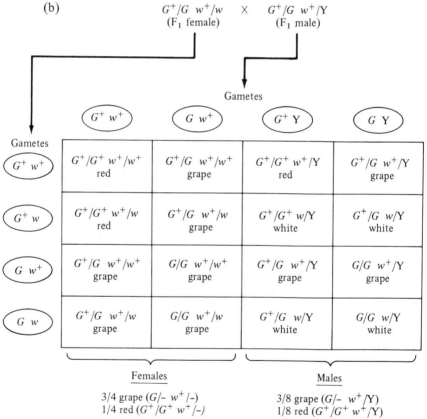

Females

    3/4 grape $(G/-\ w^+/-)$
    1/4 red $(G^+/G^+\ w^+/-)$

Males

    3/8 grape $(G/-\ w^+/Y)$
    1/8 red $(G^+/G^+\ w^+/Y)$
    1/2 white $(-/-\ w/Y)$

12-41. (a) $cn/cn\ v^+/v^+$ ♀ × $cn^+/cn^+\ v/Y$ ♂ → F_1 : $cn^+/cn\ v^+/v$ ♀ (wild type) and
$cn^+/cn\ v^+/Y$ ♂ (wild type).

(b)

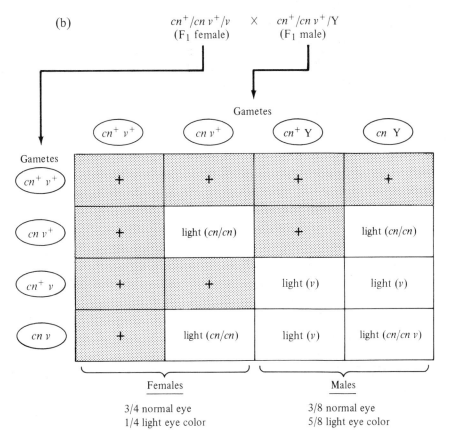

Females
3/4 normal eye
1/4 light eye color

Males
3/8 normal eye
5/8 light eye color

12-42. One of the genes (white eye) is a sex-linked recessive since none of the F_2 females are white-eyed, but half the F_2 males are white-eyed. (Males get their X chromosome only from their mothers, and the F_1 female is heterozygous for white. None of the F_2 females is white-eyed because they receive, in this case, a normal wild-type X chromosome from their fathers.) Pink is an autosomal recessive gene since both F_2 males and females show 1/4 pink among the nonwhite phenotypes. That is, although the white effect is epistatic to pink (as seen in the F_2 males), the red:pink ratio remains 3:1 since the white effect is also epistatic to red.

The crosses can be outlined as follows:

$$P_1: p/p \ w^+/w^+ \ ♀ \quad X \quad p^+/p^+ \ w/Y \ ♂$$

$$F_1: p^+/p \ w^+/w \ ♂ \quad X \quad p^+/p \ w^+/Y \ ♂$$

Gametes

	$p^+ \ w^+$	$p \ w^+$	$p^+ \ Y$	$p \ Y$
Gametes				
$p^+ \ w^+$	$p^+/p^+ \ w^+/w^+$ red	$p^+/p \ w^+/w^+$ red	$p^+/p^+ \ w^+/Y$ red	$p^+/p \ w^+/Y$ red
$p^+ \ w$	$p^+/p^+ \ w^+/w$ red	$p^+/p \ w^+/w$ red	$p^+/p^+ \ w/Y$ white	$p^+/p \ w/Y$ white
$p \ w^+$	$p^+/p \ w^+/w^+$ red	$p/p \ w^+/w^+$ pink	$p^+/p \ w^+/Y$ red	$p/p \ w^+/Y$ pink
$p \ w$	$p^+/p \ w^+/w$ red	$p/p \ w^+/w$ pink	$p^+/p \ w/Y$ white	$p/p \ w/Y$ white

Females
3/4 red
1/4 pink

Males
3/8 red
1/8 pink
1/2 white

12-43. The *Curly* gene is a dominant gene with recessive lethal effect as evidenced by the fact that it is always heterozygous and curly X curly crosses produce 2 curly:1 normal (see also text p. 190). The *Minute* gene, in this case, is a sex-linked dominant gene, lethal in hemizygous condition as evidenced by the fact that a minute X normal cross produces females, half of which show the effect, and males, none of which shows the effect (see also text p. 431).

The cross can be diagrammed as follows:

$$Cy^+/Cy\ M^+/M\ ♀ \quad × \quad Cy^+/Cy\ M^+/Y\ ♂$$

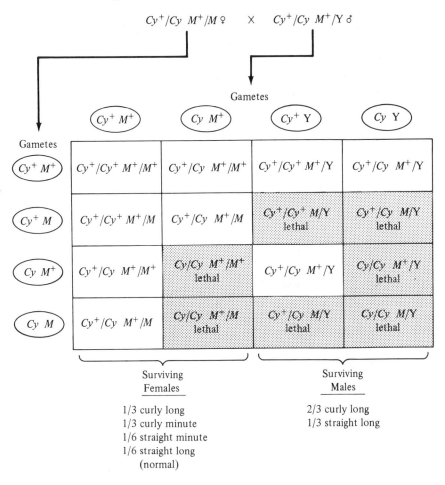

Gametes

	$Cy^+\ M^+$	$Cy\ M^+$	$Cy^+\ Y$	$Cy\ Y$
Gametes				
$Cy^+\ M^+$	$Cy^+/Cy^+\ M^+/M^+$	$Cy^+/Cy\ M^+/M^+$	$Cy^+/Cy^+\ M^+/Y$	$Cy^+/Cy\ M^+/Y$
$Cy^+\ M$	$Cy^+/Cy^+\ M^+/M$	$Cy^+/Cy\ M^+/M$	$Cy^+/Cy^+\ M/Y$ lethal	$Cy^+/Cy\ M/Y$ lethal
$Cy\ M^+$	$Cy^+/Cy\ M^+/M^+$	$Cy/Cy\ M^+/M^+$ lethal	$Cy^+/Cy\ M^+/Y$	$Cy/Cy\ M^+/Y$ lethal
$Cy\ M$	$Cy^+/Cy\ M^+/M$	$Cy/Cy\ M^+/M$ lethal	$Cy^+/Cy\ M/Y$ lethal	$Cy/Cy\ M/Y$ lethal

Surviving Females

Surviving Males

1/3 curly long
1/3 curly minute
1/6 straight minute
1/6 straight long
(normal)

2/3 curly long
1/3 straight long

12-44. Female (a) is heterozygous for the sex-linked gene *white*, w^+/w, since she produces about one half white-eyed and one half wild-type males. Female (b) is also heterozygous for this gene but is carrying, in addition, a recessive lethal gene (*l*) on its wild-type X chromosome, w^+l/wl^+. Thus, the expected wild-type male offspring of female (b) do not appear because of lethality, as evidenced by the reduced number of males.

12-45. If the cross is made $Cl\,B/sn\ ♀ × sn\ ♂$, the fertile and viable progeny are only $Cl\,B/sn$ females and sn males.

12-46. Four possible explanations are as follows:
   (a) These are *sex-limited* traits like the finding of horns only in males of some species of sheep, although all members of the species carry the genes for horns.

(b) A *Y-linked dominant gene* is responsible; that is, the Y chromosome carries the *white* dominant gene, and the X chromosome carries its recessive allele for *red* color. Thus, all females (XX) are red, and all males (XY) are white.

(c) The females bear an *attached-X chromosome* homozygous for *red* that they pass on only to their daughters and a Y chromosome that they pass on to their sons (see text pp. 210–211. The sons (XY) thus inherit their father's X chromosome, which carries *white.*

(d) Like *singed* lethality in *Drosophila* (see Problem 12–45), the trait is a sex-linked lethal and only manifests itself in the double-dose condition (XX female), not in single dose (XY male).

12-47. (a) This is supported by the fact that many genes have been found to have deleterious effects in hemizygous condition but not in heterozygous condition. Also, the effects of deleterious sex-linked genes are expected much more frequently in the heterogametic sex than in the homogametic sex, since such effects only appear in the latter when the gene is in homozygous condition (see also text p. 680).

(Reference: J.B.S. Haldane, 1922, *J. Genet.*, **12**:101–109.)

(b) One possible way to circumvent this rule is to find sex-linked genes such as the *singed* allele of Problem 12–45, which produce sterility in homozygous individuals but do not seriously affect hemizygotes. (See also a paper by W. Drescher, 1964, *Am. Nat.*, **98**:167–171.)

13

Maternal Effects and Cytoplasmic Heredity

The material in this chapter essentially discriminates between three funda-mental effects that can be presented briefly as follows:

1. *Nuclear gene.* The presence of a nuclear gene is mostly recognized by the fact that it acts in mendelian fashion. That is, it does not follow purely maternal transmission, but gives mendelian ratios in crosses: for example, an $F_1 \times F_1$ cross will produce an F_2 in the ratio of 3:1 or 9:3:3:1, etc.

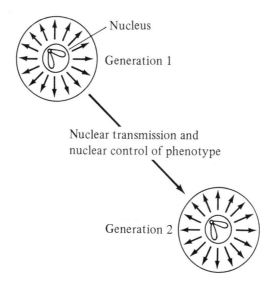

Nucleus

Generation 1

Nuclear transmission and
nuclear control of phenotype

Generation 2

2. *Extranuclear gene.* The presence of an extranuclear gene is mostly recognized by its inheritance each generation only through the maternal cytoplasm (e.g., female egg, or *Chlamydomonas* "plus" parent). Such genes are rarely, if ever, inherited through the gametes of the male parent (sperm, pollen), and they do not segregate in mendelian patterns as do nuclear genes.

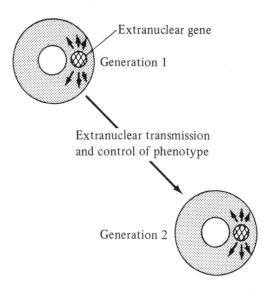

3. *Maternal effect.* Such effects are caused by nuclear genes that affect the cytoplasm of gametes (e.g., eggs). As a result, the phenotype of the zygote produced by such gametes may be different from that expected if we considered only the genotype of the gametes or zygotes themselves. Such cytoplasmic effects are chromosomally caused and last only until a new chromosome constitution occurs, which may be no more than a single generation.

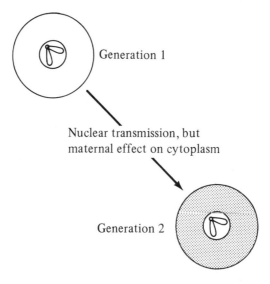

13-1. Make reciprocal crosses: (1) $T♀ × + ♂$; (2) $T♂ × + ♀$. If T is cytoplasmically inherited, there will be a difference in phenotype between the offspring of these crosses (i.e., cross 1 will produce all T phenotypes, and cross 2

will produce all + phenotypes). If T is a maternal effect, either it may wear off as development proceeds in those organisms that do not contain the T gene (e.g., larval pigment in *Ephestia kuhniella*, see text pp. 223–224), or it may last throughout the life of the organism but be corrected by a new maternal genotype in the next generation (e.g., coiling in snails, see text p. 224). Environmental determination of the trait can be judged by breeding the same genotype in different environments and observing whether the trait appears or by breeding the different phenotypic strains in the same environment and observing whether they all now develop uniformly. If inheritance of the trait does not follow any of the above patterns but segregates according to autosomal or sex-linked mendelian ratios, it can safely be assumed to be caused by chromosomal genes with nonmaternal effects.

13-2. In a cytoplasmically determined trait, the results of reciprocal crosses differ (e.g., $T ♀ \times + ♂, T ♂ \times + ♀$), but both sexes among the offspring of each cross show the same phenotype. A sex-linked trait will also produce differences between reciprocal crosses, but there will be a difference between the phenotypes of the sexes in that cross in which the homogametic parent is homozygous for the recessive allele and the heterogametic parent carries the dominant allele. [For example, $t/t ♀ \times + ♂ → t/+ ♀$ (+ phenotype):t ♂ (t phenotype).]

13-3. Although pollen is haploid, the shape of the pollen seems to be determined by the *genotype* of the plant producing the pollen rather than by the allele carried in the pollen. That is, the *long-pollen* allele has a dominant effect over that for *round-pollen* so that the heterozygous F_1 plants all produce long pollen. Only among the F_2 plants are round-pollen producers found, and these are apparently homozygous for the *round-pollen* allele. These data therefore indicate a *maternal effect*: the genotype of the anthers determines the shape of the pollen. If we call the *long-pollen* allele L and the *short-pollen* allele l, then the parental cross ($LL \times ll$) yields long-pollen–producing plants, Ll, since L is dominant. Only in the F_2 will some short-pollen–producing plants (ll) arise, in a ratio of 1:4.

(Reference: W. Bateson, E.R. Saunders, R.C. Punnett, 1905, Experimental studies in the physiology of heredity. *Reports to the Evolution Committee Royal Society.* **II.** Harrison & Sons, London.)

13-4. (a) Such a cross can be symbolized $Tra/+ XX \times +/+ XY$. Since the maternal Tra allele effects the egg cytoplasm, her daughters (XX) are intersexes, and her sons (XY) are sexually normal.

(b) Although some of the males are heterozygous for this *Transformer* allele, they cause no effect on egg development, and their progeny are therefore sexually normal males and females.

(Reference: A.H. Sturtevant, 1946, *Proc. Nat. Acad. Sci.*, 32:84–87.)

13-5. Assuming that the plastids are primarily derived from the egg cytoplasm, there is an apparent difference in the effect of this particular hybrid *hookeri/ muricata* nuclear gene combination on different plastids: *hookeri* plastids become yellow, whereas *muricata* plastids remain green.

(Reference: O. Renner, 1924, *Biol. Zentrbl.* **44**:309-336.)

13-6. The origin of the yellow sections was ascribed to plastids brought into the *muricata* ♀ X *hookeri* ♂ cross by the *hookeri* pollen rather than exclusively by the *muricata* egg cytoplasm. These plastids turned yellow because of the hybrid nucleus, but resumed their normal green color when combined with a *hookeri* nucleus.

13-7.

Chromosomal Gene (*y*)	Cytoplasmic Factor
(a) all green (*Yy*)	green ♀ X yellow ♂ → green; yellow ♀ X green ♂ → yellow
(b) *Yy* X *yy* → 1/2 green:1/2 yellow *Yy* X *YY* → all green	color will depend on female parent
(c) *Yy* X *Yy* → 3/4 green:1/4 yellow	same color as self-fertilized parent
(d) *Yy* X *Yy* → 3/4 green:1/4 yellow	color will depend on female parent

13-8. The information provided indicates that for a female plant to be variegated, it must be free of any dominant variegation inhibitors; that is, a variegated female must have inherited only recessive alleles for these inhibitor genes. If the male parent given in the problem transmitted no dominant autosomal inhibitors at all, then all of the female offspring should be variegated (i.e., *aabbcc* . . . ♂ X *aabbcc* . . . ♀ → all *aabbcc* . . .). On the other hand, if the male parent had been heterozygous for a dominant inhibitor at only *one* gene pair affecting variegation, then half of his female offspring would be expected to be variegated [i.e., *Aa* ♂ X *aa* ♀ → 1 *Aa* (green):1 *aa* (variegated)]. The fact that only about one quarter of the females were variegated fits the hypothesis that the male parent must have been heterozygous for *two* gene pairs. That is,

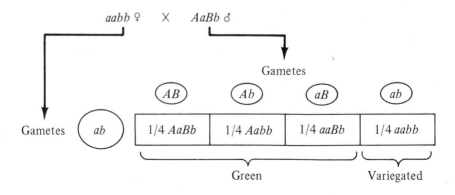

(Reference: $\emptyset$. Winge, 1931, *Hereditas*, **15**: 127–165.)

13-9. According to the problem, 127 out of 1216, or approximately one eighth of the females are variegated. Such a ratio would be expected if the cross were of the type *AaBbCc* × *aabbcc* producing offspring, seven eighths of which carry a dominant variegation inhibitor (*A*, *B*, or *C*, or a combination of these), and only one eighth carry no dominant allele (*aabbcc*). In the present case, however, the female parent is green, meaning that this plant carries one or more dominant inhibitors to variegation. Since the male parent is also green (as expected in *Lychnis* males that bear the Y chromosome), the exact parental genotypes are not discernible. Possible parental genotypes are *AaBbCc* ♀ × *aabbcc* ♂, *AaBbcc* ♀ × *aabbCc* ♂, or *Aabbcc* ♀ × *aaBbCc* ♂. Each of these crosses would produce the approximate one eighth variegated female plants.

13-10. (a) The progeny would be all streptomycin-resistant because the cytoplasmic resistance factor would be transmitted by the plus parent.

(b) In this case, the plus parent is streptomycin-sensitive, and resistance is transmitted only through a chromosomal gene of the minus parent. A single gene difference segregating in a cross between the two strains should yield zoospores in the ratio of 1/2 resistant:1/2 sensitive.

13-11. Note that a contribution of the *ss* cytoplasmic gene by the mt^- parent in cross 1 would allow growth of "heterozygous" zoospores on streptomycin-free agar, but such contribution would go undetected in streptomycin-containing agar. Similarly, a contribution of *sd* by the mt^- parent in cross 2 would allow the growth of zoospores on streptomycin agar but would go undetected on streptomycin-free agar. By these means, exceptional "heterozygous" zoospores were discovered, each of which then formed two types of clones: one kind *ss*, and the other *sd*.

(See R. Sager and Z. Ramanis, 1963, *Proc. Nat. Acad. Sci.*, **50**:260–268.)

13-12. (a) *Kk* × *Kk* → 3 killers (1 *KK*, 2 *Kk*):1 sensitive (*kk*)

(b) Descendants of sensitive exconjugant will be sensitive because they do not have *kappa*. Descendants of killer exconjugant will be in a ratio of 3 killers:1 sensitive.

(c) All sensitive.

(d) All descendants of *KK* sensitive exconjugant will be sensitive because they do not have *kappa*. All descendants of *Kk* killer will be killer because they all have *kappa* and carry at least one dose of the dominant gene *K*.

13-13. The abnormal embryos appear to be produced by a maternal effect in *o/o* ovaries.

(Reference: R.R. Humphrey, 1966, *Develop. Biol.*, **13**:57–76.)

13-14. Male sterility in this case seems to be produced by a cytoplasmic factor uninfluenced by chromosomal genes. The rare pollen produced by the male-sterile plants do not carry the cytoplasmic factor.

(Reference: M. M. Rhoades, 1933, *J. Genet.*, **37**:71-93.)

13-15. For the extranuclear factors carried in the egg cytoplasm, we can designate the male-sterile factor as *S* and the normal male-fertile factor as *s*. The chromosomal *Restorer* gene can be designated as *R* and its nonrestorer allele as *r*.
> (a) *RR* ♂ × *rr S* ♀ → all male fertile progeny (*Rr S*) since *R*– suppresses the effect of *S*.
> (b) *rr* ♂ × *Rr S* ♀ → 1/2 male fertile (*Rr S*):1/2 male sterile (*rr S*).
> *rr s* ♀ × *Rr* ♂ → all male fertile.
> (c) *Rr* ♂ × *Rr S* ♀ → 3/4 male fertile (*R*– *S*):1/4 male sterile (*rr S*).

13-16. A *simulans* X chromosome is necessary for hybrid survival in *melanogaster* cytoplasm. In *simulans* cytoplasm, however, the presence of a *melanogaster* X chromosome is generally lethal.

(Reference: A.H. Sturtevant, 1921, *Genetics*, **5**:488-500.)

13-17. A cytoplasmic change was initially induced in the cytoplasm by these treatments, as evidenced by the fact that the modification is carried through the female line. The effect is gradually replaced by normal cytoplasmic factors and disappears. The exact nature of the effect is unknown.

(Reference: F.W. Hoffman, 1927, *Genetics*, **12**:284-294.)

13-18. Loss of the *bicaudal* (*bcd*) allele can be prevented by developing a "balanced lethal" stock (see text p. 190), in which the chromosome homologous to the one carrying the *bcd* allele bears a recessive lethal allele, *le*. Such stock would produce two types of viable males, *bcd/bcd* and *bcd/le*, but only one type of fertile female, *bcd/le*, since the female homozygotes are either sterile (*bcd/bcd*) or lethal (*le/le*):

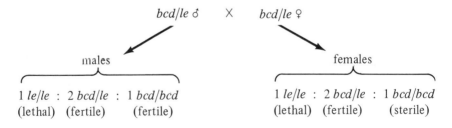

Note that no matter which type of matings occurs between any of the viable males and the fertile *bcd/le* females, the *bicaudal* allele will remain in the stock.

(Reference: A.L. Bull, 1966, *J. Exp. Zool.*, **161**:221-242.)

14
Quantitative Inheritance

14-1. (a) If the two plants are inbred in the same environment, the offspring of both plants should have similar phenotypes if plant height is environmentally caused, and show parental-type differences if plant height is genetically caused.

(b) Cross the two parental plants to obtain an F_1 and then cross $F_1 \times F_1$ to obtain an F_2. The distribution of heights in the F_2 will provide clues as to the number of genes involved in plant height. For example, a 3:1 or 1:2:1 ratio indicates one pair of genes, a 1:4:6:4:1 ratio indicates two pairs of genes, etc. As the number of gene pairs involved increases, the variability of the F_2 increases.

14-2. According to text p. 248, the number of gene pairs determining a particular quantitative trait is approximately n, where n is the exponent in the expression $(1/4)^n$. The $(1/4)^n$ value derives from the proportion of F_2 individuals resembling one of the initial homozygous parents. In the present example, probably four gene pairs are therefore involved since the proportion of F_2 plants bearing one of the initial parental phenotypes is $4/1000 = 1/250 \cong (1/4)^4$.

14-3. $(1/4)^5 = 1/1024$

14-4. (a) The gene effects appear to be additive since the F_1 is intermediate to both parental stocks and a variety of intermediate phenotypes are formed in the F_2.

(b) About three pairs, since the F_2 class resembling one of the initial parents has a frequency of $14/975 \cong 1/64 = (1/4)^3$.

(Reference: J. Clausen and W.M. Hiesey, 1958, *Carneg. Inst. Wash. Publ.*, **615**.)

14-5. (a) If we consider that each of the parental strains is probably homozygous, then the F_1, although heterozygous, is probably quite genetically uniform in the sense that all or many F_1 plants are similar in genotype. In contrast, the F_2 has an array of genotypes covering all the many genetic combinations possible between different F_1 gametes. The more gene pairs that are segregating, the greater the variation in F_2 genotypes.

(b) The parental strains may have similar ranges because they are homozygous for a similar number of additive alleles, although at different genes, for example, *AAbbCCdd* and *aaBBccDD*. The F_2, however, ranges from *AABBCCDD* to *aabbccdd*, thereby showing phenotypic effects not present in the initial parental strains.

(c) If the extreme leaf-numbered plants can be inbred, it may be possible to establish strains maintaining such extreme leaf numbers since it is likely, as explained previously, that such extreme plants are homozygous.

(d) This would not be expected if the extreme leaf-numbered plants are homozygous.

(Reference: H.K. Hayes, E.M. East, and E.G. Beinhart, 1913, *Conn. Agr. Exp. Sta. Bull.*, **176**.)

14-6. (a) Yes.

(b) Yes.

(c) No, on the assumption that light skin represents fully recessive individuals (e.g., *aabb*) who do not have genes for skin pigment (e.g., *A* or *B*).

14-7. There is apparently some dominance involved among these genes because the F_1 is more like one parent than intermediate to both parents. The wide distribution of F_2 phenotypes indicates there are a number of genes segregating for this trait, some of them acting additively. Since the frequency of the Sebright parental types is about $1/112$, we could say that three to four pairs of genes are segregating in this cross $[(1/4)^3 = 1/64, (1/4)^4 = 1/256]$. The higher frequency of F_2 phenotypes of the Hamburg type as well as the lack of an intermediate F_1 indicate that the alleles at one or more pairs of these genes may have dominant effects.

(Reference: R.C. Punnett and P.G. Bailey, 1914, *J. Genet.* 4:23-39.)

14-8. (a) 40 cm   (b) *AaBb, AAbb, aaBB*   (c) 6/16

14-9. (a) 8 cm

(b) 1 (2 cm):6 (4 cm):15 (6 cm):20 (8 cm):15 (10 cm):6 (12 cm):1 (14 cm)

(c) $(1/4)^3 + (1/4)^3 = 2/64$

(d) The F_1 height of 8 cm can be produced only by maintaining heterozygosity at no less than one gene pair (e.g., *AABbcc*). Thus, a pure-breeding 8-cm strain would not be expected to occur.

14-10. (a) 8 cm

(b) 1 (2 cm):9 (4 cm):27 (6 cm):27 (8 cm)

(c) 28/64

(d) 1/64

14-11. (a) 16 cm

(b) 1 (2 cm):6 (4 cm):15 (8 cm):20 (16 cm):15 (32 cm):6 (64 cm):1 (128 cm)

(c) 2/64

(d) The 16-cm F_1 height is determined by the presence of three dominant genes, meaning that one gene pair must always be heterozygous. Thus a 16-cm strain could not breed true.

14-12. (a) 8 cm
  (b) 1 (2 cm):7 (4 cm):18 (6 cm):22 (8 cm):13 (10 cm):3 (12 cm)
  (c) 4/64
  (d) 2/64

14-13. (a) 12 cm
  (b) 1 (2 cm):6 (4 cm):16 (6 cm):14 (10 cm):8 (12 cm):4 (14 cm):1 (16 cm)
  (c) 2/64
  (d) 2/64

14-14. (a) *AaBBCC* X *AaBBCC*   (f) *AaBbCc* X *aaBbCc*
  (b) *AaBbCC* X *AaBbCC*   (g) *aaBBCC* X *AAbbCC*
  (c) *AaBBCC* X *aaBBCC*   (h) *aaBbCC* X *AabbCC*
  (d) *AaBbCC* X *aaBbCC*   (i) *aaBbCc* X *AabbCc*
  (e) *AaBBCC* X *AabbCC*   (j) *Aabbcc* X *aaBBcc*

15

Analysis of Quantitative Characters

15-1. (a) 10.5

(b) $\sqrt{46.5/9} = 2.27$

(c) $\sqrt{5.17/10} = .719$

(d) 9.1 to 11.9

15-2. (a) 2.5

(b) 95

(c) $100 - (50 + 2.5) = 47.5$

(d) 13.5

(e) 0.5

15-3. (a) A = 10 percent; B = 15 percent; C = 20 percent

(b) Multiplicative, since the stocks with higher means show an increase in variance.

15-4. $t = \dfrac{\overline{A} - \overline{B}}{s_{\overline{A} - \overline{B}}} = -\dfrac{2.50}{1.07} = -2.33$. At 18 degrees of freedom and at the

5 percent level this t value indicates a significant difference between the two strains. (Positive and negative values of t are treated similarly.)

15-5. $t = \dfrac{16 - 28}{\sqrt{2.05}} = -8.38$. At 38 degrees of freedom this t value is highly

significant at even the .001 level.

15-6. (a) Using the method given on text pp. 260-261, the correlation coefficient r, is .76. This gives a t value of 238.69, which can certainly be considered to indicate a highly significant correlation between education and intelligence for this sample.

(b) Using Y as the difference in IQ and X as the educational difference, the regression equation calculated from these data (see text p. 262) is Y = .08 + .64X. For a zero difference in education, the expected difference in IQ between identical twins therefore would be .08. (The reference given is to the study of Newman, Freeman, and Holzinger cited in Chapter 10.)

15-7. (a) $b = .34$

(b) The regression equation is $Y = .12 + .34X$. Therefore, $X = 23$, $Y = 7.94$.

(c) $Y = 4.2$

(Reference: F. Galton, 1889. *Natural Inheritance*. Macmillan, London.)

15-8.

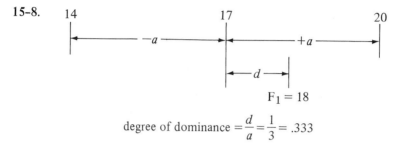

$$\text{degree of dominance} = \frac{d}{a} = \frac{1}{3} = .333$$

15-9. (a) Random breeding $= V_G + V_E = 0.366$

$$F_1 = \qquad V_E = \underline{\ .186}$$

$$\text{therefore } V_G = \ .180$$

(b) $.180/.366 = 49\%$ of total variance is attributable to genetic variance.

15-10. Environmental variance $= \dfrac{V_{P_1} + V_{P_2} + V_{F_1}}{3} = 8.864$

$$V_{F_2} = 1/2\,A + 1/4\,D + E - E = 1/2\,A + 1/4\,D = 40.350 - 8.864 = 31.486$$

$$V_{B_1} + V_{B_2} = 1/2A + 1/2D + 2E - 2E = 1/2A + 1/2D = 51.640 - 17.728 = 33.192$$

Solve as simultaneous equations by doubling both sides of the top equation and subtracting the bottom one:

$$A + 1/2\,D = 62.972$$
$$\underline{1/2\,A + 1/2\,D = 33.912}$$
$$1/2\,A = 29.060$$
$$A = 58.120$$
$$D = 9.704$$

Therefore,

(a) $d/a = \sqrt{\dfrac{D}{A}} = \sqrt{\dfrac{9.704}{58.120}} = \sqrt{.167} = .409$

(b) $h^2 = \dfrac{V_A}{V_P} = \dfrac{1/2\,A}{40.350} = \dfrac{29.060}{40.350} = 72.02\%$

(Reference: R.W. Allard, 1960, *Principles of Plant Breeding*. John Wiley & Sons, New York. See Table 10-3.)

15-11. Heritability for shank length is $\dfrac{V_A}{V_P} = \dfrac{103.4}{310.2} = 33\%$, and heritability for

neck length is $\dfrac{182.6}{730.4} = 25\%$. Since heritability for shank length is greater, this

trait therefore can be considered to be more easily changed by selection.

15-12. The effect of common genotype on the appearance of a trait observed in identical twins (compared, for example, to fraternal twins) may be quite high since this genetic effect includes dominance and epistatic effects as well as additive effects. On the other hand, heritability estimates determined for the same trait from additive effects in parent-offspring resemblances may be relatively low since dominance and epistatic effects may be excluded in appropriate experimental designs. Also, identical twins in many instances share more of a common environment than any other pair of relatives, so that the environmental effect on the phenotypic variance of identicals is relatively low. This would tend to emphasize the genetic effect on the appearance of a trait (see text pp. 170–172).

16

Linkage and Recombination

16-1. Note that the detection of recombinants depends on backcrossing the double heterozygous female to a double recessive male, because the progeny of such a cross will show all combinations of genes produced by the female. Since Sb^+ is recessive this means that the double recessive in the backcross should be $Sb^+ cu/Sb^+ cu$.

16-2. The crosses can be written as

$$\textit{bumpy tinted } \male \times Cy/Pm \; H/Sb \; \female$$

$$F_1 \; Cy \; Sb \; \male \times \textit{bumpy tinted } \female$$
$$\text{(backcross)}$$
$$\downarrow$$
$$\text{backcross progeny}$$

(a) Using text Fig. 16-8, we can note that if *bumpy* were on the second chromosome, then backcross progeny that show *Curly* would not also show *bumpy*, since they are heterozygotes and *bumpy* is a recessive. Similarly, if *bumpy* were on the third chromosome, backcross progeny that show *Stubble* would not also show *bumpy*. Therefore, since it is only the *Stubble* flies that do not show *bumpy*, *bumpy* must be on the third chromosome.

(b) In respect to *tinted*, note that tinted flies appear *both* with *Curly* and with *Stubble* among the backcross progeny. Therefore *tinted* is not on either the second or third chromosomes. However, note also that the only flies showing *tinted* are males; in fact, all of the males show *tinted*, but none of the females. This can most readily be explained if *tinted* is a sex-linked recessive. The F_1 males are therefore carrying the wild-type allele for *tinted* on their X, whereas the *bumpy tinted* females to whom they are mated are carrying *tinted* on both X chromosomes. Thus all males among the backcross progeny are tinted and all females are wild-type heterozygotes.

16-3. *Adh* is located on the second chromosome because the slow form of the enzyme, contributed by the wild-type stock, always segregates from *Curly*, the second chromosome marker carrying the fast form, that is, $Cy/+ = Adh^{Fast}/Adh^{Slow}$, $+/+.= Adh^{Slow}/Adh^{Slow}$.

(Reference: E.H. Grell, K.B. Jacobson, and J.B. Murphy, 1965, *Science*, **149**:80-82.)

16–4. If *short* (*s*) were on the second chromosome then the F_1 males can be described as having the chromosome constitution (*black* = *b*, *pink* = *p*):

	Parental (noncrossover)		Recombinant (crossover)	
	b s p		*b + p*	
	b s +		*b + +*	
	+ + p		*+ s p*	
	+ + +		*+ s +*	

2nd: *b s* + +   3rd: *p* / +

Gametes →

A backcross of these males to homozygous *black short pink* females would then be expected to produce more parental phenotypic combinations among the offspring (black short pink, black short +, + + +, + + pink) than recombinants (black + pink, black + +, + short pink, + short +). The results, however, show no indication of linkage between *black* and *short* since there are 204 parental types and 203 recombinant types among the progeny. Similarly, if *short* were on the third chromosome and the F_1 males were

2nd: *b* / +   3rd: *p s* / + +

more parental combinations for this chromosome would be expected (black pink short, black + +, + + +, + pink short) than recombinants (black pink +, black + short, + pink +, + + short). Again, the results do not demonstrate linkage between *short* and *pink* because there are 198 parentals and 209 recombinants.

If *short* were on the X chromosome, the F_1 males could be represented as

X (1st): *s* / Y   2nd: *b* / +   3rd: *p* / +

Thus, if *s* were sex-linked, all of the backcross progeny would be short because the female parent is homozygous *short*. This is obviously not so.

By elimination of the other alternatives, the answer is therefore that *short* is located on a small fourth chromosome and assorts independently of *black*, *pink*, or the sex chromosome. That is, the F_1 male is

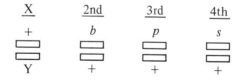

(Reference: H.J. Muller, 1914, *J. Exp. Zool.*, **17**:325–336.)

16-5.

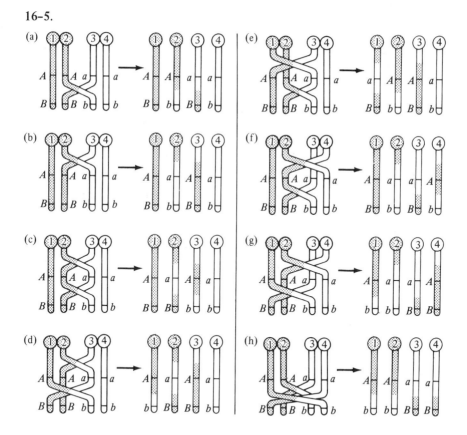

16-6. Marriages should be examined in which either independent assortment or linkage can be detected among the gametes produced by one of the marriage partners. For example, using a simplified notation for the blood group loci, the double heterozygote in the mating *AB MN* × *OO NN* should (under independent assortment) produce four kinds of gametes in equal proportions (*AM, AN, BM, BN*), leading to the production of four types of offspring in the ratio 1 *AO NN*:1 *AO MN*:1 *BO NN*:1 *BO MN*. If linkage existed between the two gene pairs, then the cross of phenotypes AB MN $\times$ O N might be diagrammed as

In the first case, the parental combinations among the progeny would produce phenotypes A MN, B N, and the recombinant combinations would be A N, B MN; whereas in the second case, the phenotypic combinations would be reversed. In either case, more parental combinations than recombinants would be expected in the case of linkage, and approximately equal numbers are expected in the case of independent assortment.

16-7. $(638 + 672)/(21,379 + 21,096 + 638 + 672) = 1,310/43,785 = 2.99\%$.

(Reference: C.B. Hutchison, 1922, *Cornell Agr. Exp. Sta. Mem.*, **60**.)

16-8. 95 percent confidence interval = observed ratio $\pm 1.96 \sqrt{pq/N} = .0299 \pm 1.96 \sqrt{(.03)(.97)/43,785} = .0299 \pm .0016$. This gives a confidence interval from .0283 to .0315.

16-9. To solve this problem see text p. 283 and Fig. 16-8. The recessive gene *green* must be located on the second chromosome since it never appears phenotypically together with *Curly* in the final cross given. *Jumpy* is also a recessive trait that must, however, be located on the fourth chromosome because it can appear phenotypically either with *Curly* or with *Stubble*. (*Jumpy* is not sex-linked since, if it were, the F$_1$ males would not be carrying the gene, and the jumpy phenotypes would appear only among the male progeny of the backcross but not among the females.)

16-10. Since XXY females produce XX nondisjunctional eggs which then become XXY females (when fertilized by a Y-bearing sperm), such female offspring may show the parental combination on one X chromosome and the recombinant combination on the other X because of 4-strand crossing over. An example of this is shown in the figure at the top of the next page. Such individuals have been found.

16-11. Genes for these traits can arbitrarily be designated as S = spotted, s = self-colored, A = short-haired, a = angora. The parental cross is therefore $SS\,AA \times ss\,aa$, and the backcross can be diagrammed as

$$\frac{S\,A}{s\,a} \times \frac{s\,a}{s\,a}$$

Figure for Problem 16-10

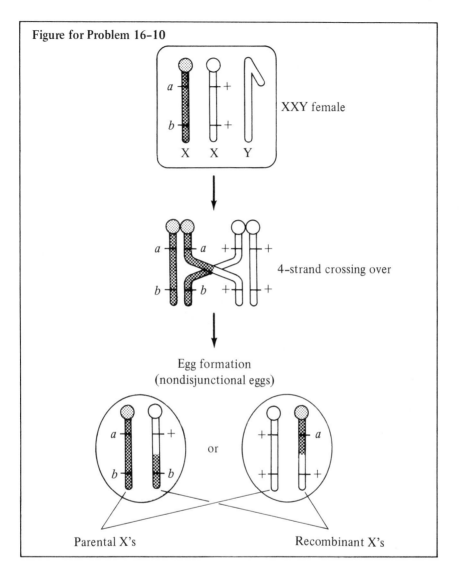

XXY female

4-strand crossing over

Egg formation
(nondisjunctional eggs)

or

Parental X's   Recombinant X's

The recombinant phenotypes among the progeny are therefore S a (spotted angora) and s A (self-colored, short-haired). Their frequency is $(26 + 23)/(26 + 144 + 157 + 23) = 49/350 = 14.0\%$.

16-12. (a) Plant $1 = R\ Y/r\ y$; plant $2 = R\ y/r\ Y$

(b) 10.0 percent

(c) Since r (recombination frequency) $= .10$, plant 1 produces $r\ y$ gametes in frequency of $1/2 - 1/2r = .45$ (see text Table 16-2). Plant 2 produces $r\ y$ gametes in frequency of $1/2r = .05$. The frequency of $r\ y/r\ y$ zygotes in a cross between the two plants is therefore $.45 \times .05$, or 2.25 percent.

16-13. Based on the information given in the problem, the F_1 can be symbolized fS/Fs, with fS on one homologue and Fs on the other. Since the recombination frequency between the two loci is .15, this means that a frequency of .85 parental gametes are produced by the F_1 or approximately .425 fS and .425 Fs, along with .15 recombinant gametes, or .075 FS and .075 fs. The answers are therefore as follows:

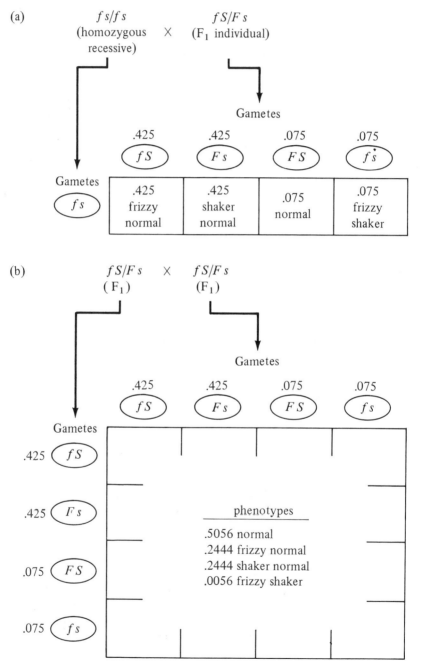

(a)

<table>
<tr><td colspan="2"></td><td>fs/fs</td><td></td><td>fS/Fs</td></tr>
<tr><td colspan="2"></td><td>(homozygous</td><td>X</td><td>(F_1 individual)</td></tr>
<tr><td colspan="2"></td><td>recessive)</td><td></td><td></td></tr>
</table>

Gametes

	.425	.425	.075	.075
	fS	Fs	FS	fs

Gametes fs

.425 frizzy normal	.425 shaker normal	.075 normal	.075 frizzy shaker

(b)

fS/Fs X fS/Fs
(F_1)   (F_1)

Gametes

.425	.425	.075	.075
fS	Fs	FS	fs

Gametes

.425 fS

.425 Fs

.075 FS

.075 fs

phenotypes

.5056 normal
.2444 frizzy normal
.2444 shaker normal
.0056 frizzy shaker

16-14. (1) Lethality: Note that the recessive phenotypes, a and b, are less frequent than expected (expected A:a = 200:200; observed A:a = 220:180. Expected B:b = 200:200, observed B:b = 260:140). Note also that linkage is not responsible since the observed numbers of parental phenotypes (A b + a B = 200) is the same as the observed numbers of recombinants (A B + a b = 200).

    (2) In this case, linkage appears to be responsible for the departure from independent assortment since no lethality among either a or b phenotypes can be detected (observed A:a ratio = 200:200, and B:b ratio = 200:200). The number of recombinants (A B + a b = 100) is significantly smaller than the number of parentals (A b + a B = 300) indicating linkage.

16-15. (a) Linkage cannot be detected because the products of crossing over (e.g., *A b* and *a b* gametes) are identical to the parental noncrossover gametic products (e.g., *A b* and *a b*).

    (b) In this case, linkage can be detected by noting departures from the ratios expected under independent assortment. Note that if the two gene pairs assort independently, the *Aa Bb* parent produces four types of gametes with equal frequency, whereas the *Aa bb* parent produces only two. The offspring should therefore appear in the following *phenotypic* frequency:

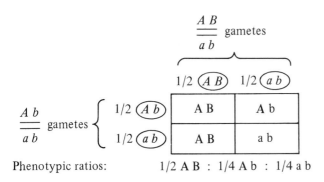

However, if the two genes are linked, the frequency of these phenotypes will change. For example, if the genes are completely linked with *A B* on one chromosome and *a b* on its homologue, then aB phenotypes will be absent:

$$\frac{A\ B}{a\ b}\ \text{gametes}$$

	1/2 Ⓐ Ⓑ	1/2 Ⓐ Ⓑ
1/2 Ⓐ Ⓑ	A B	A b
1/2 Ⓐ Ⓑ	A B	a b

Phenotypic ratios: 1/2 A B : 1/4 A b : 1/4 a b

If there is incomplete linkage between the two gene pairs, some aB phenotypes will appear but their frequency will be less than that of the ab offspring
and less than the one-eighth frequency expected under independent assortment.
For example, a 10 percent recombination frequency between the two gene pairs
will produce the following:

	90% parentals		10% recombinants	
	.450 (A B)	.450 (a b)	.050 (A b)	.050 (a B)
.500 (A b)	.225 A B	.225 A b	.025 A b	.025 A B
.500 (a b)	.225 A B	.225 a b	.025 A b	.025 a B

Phenotypic ratios: .475 A B : .275 A b : .225 a b : .025 a B

(Would the ability to detect linkage change if the *Aa Bb* parent had its genes
linked in repulsion with *A b* on one chromosome and *a B* on its homologue?)

16-16. (a) *A b/a b*   (d) *a B/a b*
     (b) *A B/a b*   (e) *A b/a B*
     (c) *A B/A b*   (f) *A B/a B*

16-17. A chi-square test for independent assortment (text pp. 133-134 and
Table 16-1c) yields $\chi^2 = 3.45$, which, at 1 degree of freedom and the 5 percent
level of significance, allows the hypothesis of independent assortment to be accepted.

(Reference: R.A. Fisher and G.D. Snell, 1948, *Heredity*, 2:271-273.)

16-18. The departure from independent assortment in this case seems to be caused
by both abnormal segregation and linkage. Chi-square tests for 3:1 ratios show
a higher-than-expected ratio of wild type to oligodactyly ($\chi^2_{1\,df} = 13.29$) and a
lower-than-expected ratio of wild type to albino ($\chi^2_{1\,df} = 13.29$). A chi-square
test for independent assortment yields a value of 27.61, which enables us to
reject the hypothesis of independent assortment and indicates the presence of
linkage.

(Reference: P. Hertwig, 1942, *Zeit, Ind. Abst. und Vererbung.*, 80:220-246.)

16-19. The cross can be visualized as

$$\frac{A\,B}{a\,b}\,\frac{C}{c} \times \frac{a\,b}{a\,b}\,\frac{c}{c}$$

(a) Since linkage is complete between the *A* and *B* genes, the F_1 heterozygote produces four types of gametes in equal proportion (*A B C, A B c, a b C,*

a b c) and the phenotypes expected are therefore 1/4 A B C:1/4 A B c:1/4 a b C: 1/4 a b c.

(b) When recombination frequency between *A* and *B* is 20 percent, the F_1 heterozygote will produce 80 percent parental gametes for these genes (e.g., 40 percent *A B*, 40 percent *a b*) and 20 percent recombinants (e.g., 10 percent *A b*, 10 percent *a B*). Because each of these combinations will assort independently in respect to *C* and *c*, about half of each of these combinations will be *C* and the other half *c* (e.g., 40 percent *A B* = 20 percent *A B C* + 20 percent *A B c*). The expected F_2 phenotypes will therefore be

20 percent A B C:20 percent A B c:20 percent a b C:20 percent a b c: 5 percent A b C:5 percent A b c:5 percent a B C:5 percent a B c

16-20. (a) The heterozygous F_1 females are so constructed that *A B C* enters from one parent, and *a b c* from the other. On this basis, the progeny of the backcross can be classified as to whether they are parental or recombinant for each pair of genes by observing whether the parental phenotypic combinations are retained (e.g., A - B, a - b) or changed (e.g., A - b, a - B).

	Backcross Phenotypes	
Gene Pair	Parentals	Recombinants
A - B	a b c + A B C = 420	a B c + A b C = 420
A - C	a b c + A B C + a B c + A b C = 840	none
B - C	a b c + A B C = 420	a B c + A b C = 420

These data indicate that there is independent assortment between *A - B* and between *B - C*, but that genes *A* and *C* are so closely linked that no recombinants occur. If all three loci had been segregating independently, the F_2 ratio would have been 1:1:1:1:1:1:1:1 between the eight different possible phenotypes.

(b) 0 percent

16-21. In this case, the F_1 females are genotypically *DEF/def* and the backcross progeny can be divided into the following phenotypic groups:

	Backcross Phenotypes	
Gene Pair	Parentals	Recombinants
D - E	d e f + D E F + d e F + D E f = 100	d E F + D e f + d E f + D e F = 100
E - F	d e f + D E F + d E F + D e f = 180	d e F + D E f + d E f + D e F = 20
D - F	d e F + D E F + d E f + D e F = 100	d E F + D e f + d e F + D E f = 100

(a) Based on the above analysis, there is no apparent linkage between D-E and D-F, but E-F does show linkage.

(b) $20/200 = 10$ percent

16-22. (a) The male produces only the nonrecombinant $A\ b$ and $a\ B$ gametes whereas the female produces both nonrecombinant ($A\ b$ and $a\ B$) and recombinant ($A\ B$ and $a\ b$) gametes. Note that the zygotes produced will be in the *phenotypic* ratio of 2 A B:1 Ab:1 a B whether recombination has or has not taken place:

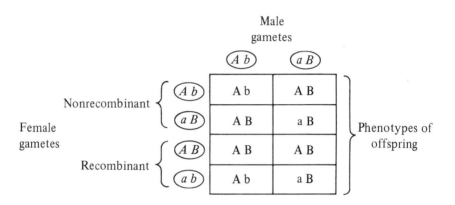

(b) Although the extent of recombination between the A and B genes would affect the frequency of $A\ B$ and $a\ b$ gametes produced by the female, it would obviously not alter the phenotypic ratios among the offspring, since fertilized recombinant gametes produce the same ratios as fertilized nonrecombinant gametes.

16-23. (a) Make the cross $aaBB \times AAbb$ to obtain an F_1 of genotype $a\ B/A\ b$.

Cross F_1 males and females $\dfrac{a\ B}{A\ b} \times \dfrac{a\ B}{A\ b}$ to obtain an F_2. Some of the F_2 a phenotypes will be heterozygous for b ($a\ B/a\ b$) because of crossing over in the F_1 females. (Since there is no crossing over in male *Drosophila*, all a phenotypes will be carrying the B from their fathers.) If these F_2 a phenotypes are crossed with each other, a number of these matings therefore will be of the type $a\ b/a\ B \times a\ b/a\ B$ and will produce $aabb$ offspring in a ratio of 25 percent.

(b) Make the cross $a\ B/a\ B\ B\ ♀ \times A\ b\ b\ ♂$ to obtain F_1 females of genotype $a\ B/A\ b$. Crossovers will occur in these females who will produce sons of ab genotype and phenotype. These $a\ b$ males can be backcrossed to the F_1 females and the offspring of this cross observed for the presence of $aabb$ daughters who have arisen as a result of crossing over.

16-24. Since there is no crossing over in *Drosophila* males, only two types of sperm are produced, $A\ B$ and $a\ b$, whereas four types of eggs are produced:

.425 *A B*:.425 *a b*, .075 *A b*:.075 *a B*. Random combination of eggs and sperm then gives *phenotypes* in the ratio:

$$.7125 \text{ A B}:.0375 \text{ A b}:.0375 \text{ a B}:.2125 \text{ a b}$$

16–25. (a) Males and females in the phenotypic ratio .425 A B:.075 A b:.075 a B:.425 a b.

(b) Males and females in the phenotypic ratio .075 A B:.425 A b:.425 a B:.075 a b.

(c) Males as in (a); females, all A B.

(d) Males as in (a); females, 50 percent A B, 50 percent A b.

(e) Males as in (b); females as in (d).

16–26. If we assume that each strain is homozygous for a different recessive temperature-sensitive mutation (e.g., *a* +/*a* + and + *b*/+ *b*), one answer would be to cross the two strains and then perform an F_1 backcross to a double homozygous recessive (*a* +/+ *b* X *a b*/*a b*) at the permissive temperature. If the backcross offspring are raised at the restrictive temperature, only the wild-type recombinant product (+ +/*a b*) will survive:

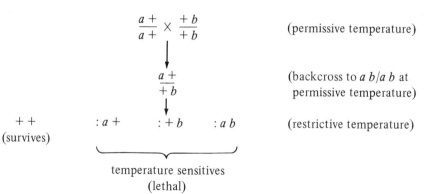

(See D.T. Suzuki, 1970, Temperature-sensitive mutations in *Drosophila melanogaster, Science,* **170**:695–706.)

17
Gene Mapping in Diploids

THREE-POINT LINKAGE ANALYSIS

The general procedure for determining gene order and linkage distances in a three-point test is briefly outlined on text pages 294-295. Using this method, analysis of an arithmetical example for three linked genes, A, B, C, might proceed as follows:

Parental cross:	$AA\ BB\ CC \times aa\ bb\ cc$
F_1 backcross:	$Aa\ Bb\ Cc \times aa\ bb\ cc$

F$_2$ phenotypic data:

① A b C = 24
② A b c = 49
③ a B C = 49
④ a B c = 24
⑤ A B c = 1
⑥ a b C = 1
⑦ A B C = 426
⑧ a b c = 426
Total: 1,000

	Recombinants		Parentals
A-B Interval	**B-C Interval**	**A-C Interval**	⑦ A B C
① A b C 24	① A b C 24	② A b c 49	⑧ a b c
② A b c 49	④ a B c 24	③ a B C 49	
③ a B C 49	⑤ A B c 1	⑤ A B c 1	
④ a B c 24	⑥ a b C 1	⑥ a b C 1	
Total: 146	Total: 50	Total: 100	
Recombination percent $= (146/1000) \times 100$ $= 14.6$	Recombination percent $= (50/1000) \times 100$ $= 5.0$	Recombination percent $= (100/1000) \times 100$ $= 10.0$	

Tentative Map:

$$B \xleftrightarrow{\quad 5.0 \quad} C \xleftrightarrow{\quad 10.0 \quad} A$$
$$\longleftarrow 14.6 \longrightarrow$$

Tentative map shows that F_1 backcross was genotypically: $\dfrac{B\,C\,A}{b\,c\,a} \times \dfrac{b\,c\,a}{b\,c\,a}$

Note, however, for linear order B-C-A,
   double crossovers between B and A are
   shown by the two classes ⑤ and ⑥

⑤ B c A 1
⑥ b C a 1
   Total: 2

or, percent double crossover individuals = (2/1000) × 100 = .2.

But each double crossover individual represents *two* crossovers between B-A, or ".2" really equals .2% × 2 = .4% crossovers in the B-A interval.

Therefore, value of singles between B-A = 14.6
   Plus value of doubles between B-A =   .4
      Total distance B-A = 15.0

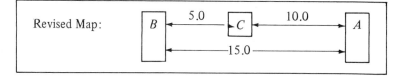

Revised Map:

17-1. (a) C-Sh distance = (509 + 524 + 20 + 12)/45,832 = 2.32 percent. Sh-Wx distance = (4455 + 4654 + 20 + 12)/45,832 = 19.95 percent. C-Wx distance = (509 + 524 + 4455 + 4654)/45,832 = 22.13 percent. The gene order is therefore C-Sh-Wx, and the double crossover classes are obviously $C\,sh\,Wx$ and $c\,Sh\,wx$. On this basis, the C-Wx distance should be increased by twice the frequency of the double crossovers [(20 + 32)/45,832 = .07] and equals 22.13 + 2 (.07) = 22.27 percent. Thus the linkage map is as follows:

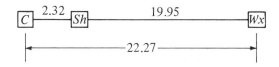

(b) coincidence $= \dfrac{\text{observed frequency double crossovers}}{\text{expected frequency double crossovers}}$

$$= \dfrac{.0007}{(.0232)(.1995)} = \dfrac{.0007}{.0046} = .15$$

interference = 1 − coincidence = 1 − .15 = .85 = 85 percent.

(Reference: L. J. Stadler, 1926, *Genetics*, **11**:1-37.)

17-2. Since the parental classes would be expected to be those that are most frequent, one parental chromosome can be considered as + + s and the other as o p + (without regard to gene order). On this basis, crossovers between the o and p loci produce phenotypes that are either o + or + p, crossovers between

o and s produce phenotypes o s or $+ +$, and crossovers between p and s produce phenotypes p s or $+ +$.

(a) Note that the double crossovers are the least frequent phenotypic classes, $+ p +$ and o $+$ s, and the origin of these classes could most easily be explained as resulting when the center gene is o, i.e., the parental chromosome combination is $s + +/+ o p$. Thus, the gene sequence is s-o-p. The sequence can also be determined by using the map distances calculated in (c).

(b) $s + +/s + +, + o p/+ o p$.

(c) o-p distance $= (2 + 96 + 110 + 2)/1000 = 21$ percent.

    o-s distance $= (73 + 2 + 2 + 63)/1000 = 14$ percent.

    p-s distance $= (73 + 96 + 110 + 63)/1000 = 34.2$ percent.

From these calculations the p-s distance obviously includes the two smaller distances, and the gene sequence can be written as p-o-s or s-o-p. Note that for this sequence the $+ p +$ and o $+$ s classes are obviously double crossovers, and the p-s distance therefore should be increased by twice the frequency of these classes, or $34.2 + 2(.4) = 35$ percent.

(d) Expected frequency of double crossovers $= .21 \times .14 = .029$.

$$\text{coincidence} = \frac{\text{obs.}}{\text{exp.}} = \frac{.004}{.029} = .14 = 14 \text{ percent}$$

17-3. (a) a-d: A d rg; a D Rg; A d Rg; a D rg

          d-rg: A D rg; a d Rg; A d Rg; a D rg

          a-rg: A D rg; a d Rg; A d rg; a D rg

(b) a-$d = 420/1000 = 42$ percent; d-$rg = 300/1000 = 30$ percent; a-$rg = 200/1000 = 20$ percent. The gene order is therefore a rg d, and twice the frequency of the double crossovers (A D rg, a d Rg) should be included in the a-d distance, or a-$d = 42 + 2(4) = 50$ percent. The linkage map is therefore:

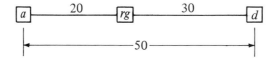

(c) coincidence $= \dfrac{\text{observed frequency doubles}}{\text{expected frequency doubles}} = \dfrac{.04}{.20 \times .30} = \dfrac{.04}{.06} = .67$

(d) interference $= 1 -$ coincidence $= 1 - .67 = .33 = 33$ percent

(e) A reduction in the frequency of double crossovers, that is, less of the A D rg and a d Rg phenotypes.

(f) 40 percent a Rg: 40 percent A rg: 10 percent a rg: 10 percent A Rg

17-4. (a) If we number the eight phenotypic classes from ① to ⑧ in the order they are given (i.e., y sh c = ① , Y Sh C = ② , etc.), the linkage distances are as follows for each plant:

Linkage Distance	Plant I	Plant II	Plant III
y - sh	①+②+⑤+⑥= 24%	③+④+⑦+⑧= 24%	①+②+⑤+⑥= 24%
sh - c	①+②+③+④= 20%	⑤+⑥+⑦+⑧= 20%	⑤+⑥+⑦+⑧= 20%
y - c	③+④+⑤+⑥= 5%	③+④+⑤+⑥= 5%	①+②+⑦+⑧= 5%

The linkage map is then:

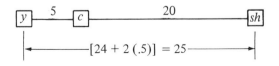

$$\frac{[24 + 2 (.5)] = 25}{}$$

(b) $\dfrac{\text{obs.}}{\text{exp.}} = \dfrac{.005}{(.200)(.050)} = \dfrac{.005}{.010} = .5 = 50 \text{ percent}$

(c) The frequency of the double-crossover classes would be expected to decrease, that is, classes ③ + ④ for plant I, classes ⑤ + ⑥ for plant II, classes ⑦ + ⑧ for plant III.

(d) Plant I = y c Sh/Y C sh; Plant II = y c sh/Y C Sh; Plant III = y C Sh/Y c sh

17-5. These alternatives would not be distinguishable unless crossing over were absent in one of the sexes, as in male *Drosophila*.

17-6.

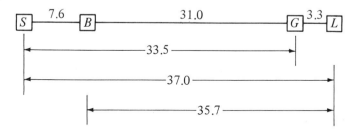

(Reference: R.P. Gregory, D. DeWinton, and W. Bateson, 1923, *J. Genet.*, 13:219-253.)

17-7. (a) 40 percent (c) A B and a b
   (b) 20 percent (d) A b and a B

17-8. (a) Since each chiasma represents a crossover event in only two out of four chromatids (see discussion text pp. 306-307), the occurrence of a single chiasma in 100 meiotic divisions will only produce 2 recombinant gametes for that interval out of a total of 400; or the chiasma frequency (1 percent) will be twice the recombinant frequency (0.5 percent). In the present problem, the given linkage distances of 10 percent and 20 percent should therefore provide chiasma frequencies of about 20 percent and 40 percent, respectively. Some of these chiasmata, however, will occur as "double chiasmata" just as some of the genetic crossovers determining linkage distances occur as double crossovers. Because one double chiasma produces only two out of four chromatids showing recombinant double crossover products, we would also expect the frequency of double chiasmata to be twice the double crossover frequency. Specifically, $.10 \times .20 = .02$ double crossovers are expected for the two intervals, and therefore .04 double chiasmata are expected.

   (b) The linkage distance between a - b is determined by adding the frequency of double crossovers occurring simultaneously in the two intervals (.02) to the frequency of single crossovers occurring in the a - b interval alone $(.10 - .02 = .08)$. Since the .08 value of the latter corresponds to twice that many single chiasmata, the a - b single chiasma frequency is .16.

   (c) Following the same reasoning as in (b), there is a b - c frequency of .36 single chiasmata $[2(.20 - .02)]$. The sum total of chiasmata is therefore .04 (doubles) + .16 (a - b singles) + .36 (b - c singles) = .56, and the proportion of cells without chiasmata is therefore $1 - .56 = .44$.

17-9. (a) $2(.10 \times .20 \times .5) = 2(.01) = .02$
   (b) $2(.10 - .01) = .18$
   (c) $1 - (.02 + .18 + .38) = 1 - .58 = .42$

17-10. The expected frequency of double crossovers is $.10 \times .20 \times .5 = .01$. Thus, although a total of 30 percent recombinants are expected in each cross, 1 percent will not be detectable because of double crossing over. Therefore,

   (a) 29 percent a + and + c recombinants:71 percent a c and + + nonrecombinants.

   (b) 29 percent a c and + + recombinants:71 percent a + and + c nonrecombinants.

17-11. (a) D - e distance $= (214 + 187 + 28 + 31)/2191 = 21.0$ percent
         D - p distance $= (11 + 18 + 28 + 31)/2191 = 4.0$ percent
         p - e distance $= (214 + 187 + 11 + 18)/2191 = 19.6$ percent
   The gene order is therefore D - p - e, and the double crossovers are D p + and + + e, or $11 + 18 = 1.3$ percent. Twice the frequency of double crossovers must then be added to the D - e distance, which is therefore $21.0 + 2(1.3) = 23.6$ percent. The linkage map is

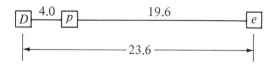

(b) The expected frequency of double crossovers is .04 $\times$ .196 = .0078, and the observed frequency is .00132. Thus, the coefficient of coincidence is .0132/.078 = 1.7. Note that there is no interference in this cross. In fact, the number of observed double crossovers is somewhat higher than expected. One important reason for this is the location of the centromere near the *pink* locus (see text Fig. 17-3). Numerous experiments have shown that there is little or no interference to double crossing over in *Drosophila* when loci are on opposite sides of the centromere (see text p. 308).

(c) .04 $\times$ .196 $\times$ 1.7 $\times$ 1000 = 13

17-12. (a) *al - d* distance = (7 + 5 + 132 + 100)/2118 = 11.5 percent
al - S distance = (7 + 5)/2118 = .6 percent
S - d distance = (132 + 100)/2118 = 10.9 percent

The gene order is therefore *al - S - d*, and the F_1 backcross can be written as

$$\frac{al^+\ S\ dp^+}{al\ \ S^+\ dp} \times \frac{al\ S^+\ dp}{al\ S^+\ dp}$$

(b) The F_2 phenotypes that are missing are the double crossovers, *al S d* and + + +. They are absent because their expected frequency, .109 $\times$ .006 = .0006, is so low that only one double crossover would be expected among about 2000 progeny. It is therefore not surprising that no double crossovers appeared in the 2118 flies counted.

(Reference: C. Stern, and C.B. Bridges, 1926, *Genetics*, **11**: 503-530.)

17-13. The least frequent phenotypes would be
(a) + b +, a + c   (c) a + +, + b c
(b) a b c, + + +   (d) + + c, a b +

17-14. In each case, the results will follow the pattern given in the answers for Problem 17-13, except that the given mutant alleles are now dominant. For example, condition I will produce double crossover phenotypes (a) + B +, A + C; (b) A B C, + + +; (c) A + +, + B C; (d) + + C, A B +. Condition II will produce double crossover phenotypes (a) + B +, A + C; (b) A B C, + + +; etc.

17-15. (a) Note that Adh^S enters with the wild-type genes for *b, el, rd*, and *pr* and stays with the wild-type genes for *el* and *rd* in every case except when there are crossing-over events between *el* and *rd*. This indicates that the *Adh* locus is most likely located between *el* and *rd*.

(b) There are 3 + 25 + 17 + 2 = 47 crossover events between *el* and *rd*, of which 5 represent crossing over between the *el* locus and *Adh* [+ Adh^S/el $Adh^F \rightarrow el\ Adh^S$ (3), + Adh^F (2)]. In other words, *Adh* is 5/47 along the distance between *el* and *rd*, or 50 + 5/47 (1.0) = 50.1 = *Adh* locus.

17–16. (a) Note that l has entered the F_1 female linked to the wild-type alleles for the four genes. If l were located between a closely linked pair of these genes (e.g., $sc^+ - cv^+$, or $cv^+ - v^+$, or $v^+ - f^+$), F_2 male zygotes that are wild-type for this pair of genes would be expected to be carrying l (except for rare double crossovers). However, since l is lethal in hemizygous condition, such males will not appear. Since the only males missing that are wild type for a closely linked pair of genes are in the $v^+ f^+$ class, this indicates that l is probably located between v and f.

(b) In the v -f interval the F_1 female can be represented as $v^+ - l - f^+ / v - l^+ - f$. Crossovers in the v - l interval therefore should produce $v^+ l^+ f$ and $v l f^+$ males, of which only the former will survive. The observed number of $v^+ f$ males is $20 + 6 = 26$, and the number of observed crossovers in the v - f interval is $43 + 20 + 7 + 1 + 6 + 11 = 88$. Thus, l is located $26/88 = .29$ along the distance from v to f, or the l locus is $33 + .29(23.7) = 39.9$.

17–17. (a) The lethal gene is probably not located between rb and ct since, if it were, the wild-type class in the male offspring could only arise by double crossing over ($+ l + / rb + ct \rightarrow + + +$), and this class is certainly much too common (e.g., compared to the $+ + ct$ class) to indicate this origin. The alternatives are that the order is l - rb - ct or rb - ct - l. In the former case, the l - rb distance should be less than the l - ct distance. In the latter case, the reverse should be true. These distances can be measured by noting that crossovers between $l + /l^+ rb$ should produce $l rb$ (lethal) and $l^+ +$ (wild-type for *ruby*) males, whereas crossovers between $l + /l^+ ct$ should produce $l ct$ (lethal) and $l^+ +$ (wild-type for *cut*) males. The observed numbers of these classes among F_2 males are $3.90 + .10 = 4.00$ percent crossovers between l - rb, and $3.90 + 12.40 = 16.30$ percent crossovers between l - ct. The linkage order is obviously l - rb - ct. Note that for this linkage order there are two possible double crossover classes, $l rb +$ (lethal) and $+ + ct$ (frequency .10 percent). Thus, the l - ct distance should be increased by twice this frequency or l - $ct = 16.30 + 2(.10) = 16.50.$*

(b) Since *ruby* occupies the locus 7.5, and ct has the locus 20.0, l must be at the 3.5 position.

(c) Since the rb - ct distance observed among males is $12.40 + .10 = 12.5$ (note that this is the same distance observed among females) and the l - rb distance is 4.0, the expected frequency of double crossovers is $.125 \times .040 = .0050$. The coincidence coefficient is therefore $.0010/.0050 = 20$ percent and the degree of interference is 80 percent.

17–18. The method used is given on text pp. 296–297, and proceeds as follows:

---

*It is possible to estimate the frequency of the lethal double-crossover class ($l rb +$) as having the same frequency as the observed viable double-crossover class ($+ + ct$). This procedure, however, would complicate the calculations since we would then have to modify the observed total number of males to account for the numbers of estimated lethal genotypes. For simplicity, therefore, the answers have been calculated on the basis of the given frequencies.

Double crossovers $(\alpha \beta \gamma/+ + + \rightarrow \alpha + \gamma, + \beta +)$:

Frequencies expected without interference: $(\alpha - \beta \text{ distance}) \times$
$(\beta - \gamma \text{ distance}) \times 1000 = .1 \times .1 \times 1000 = 10$

Frequencies expected with interference: expected double crossovers
$\times$ coincidence value $= 10 \times .6 = 6$, for example, $3\alpha + \gamma$ and
$3 + \beta +$

Single crossovers $(\alpha \beta \gamma/+ + + \rightarrow \alpha + +, + \beta \gamma, \alpha \beta +, + + \gamma)$:

$\alpha - \beta$ frequency: $.1 \times 1000 = 100 - 6 \text{ (doubles)} = 94$ singles, for
example, $47\alpha + +$ and $47 + \beta \gamma$

$\beta - \gamma$ frequency: same as $\alpha - \beta$ but different classes, for example,
$47 \alpha \beta +$ and $47 + + \gamma$

Parentals $(\alpha \beta \gamma, + + +)$:

$1000 - \text{total no. of crossovers} = 1000 - (6 + 94 + 94) = 806$,
for example, $403 \alpha \beta \gamma$ and $403 + + +$

Totals:

$$\begin{aligned}
\alpha \beta \gamma \text{ and } + + + &= 806 \\
\alpha + + \text{ and } + \beta \gamma &= 94 \\
\alpha \beta + \text{ and } + + \gamma &= 94 \\
\alpha + \gamma \text{ and } + \beta + &= \underline{6} \\
& 1000
\end{aligned}$$

17-19. For method, see answer to Problem 17–18 above:

$$\begin{aligned}
+ \beta \gamma \text{ and } \alpha + + &= 806 \text{ (parentals)} \\
\alpha \beta \gamma \text{ and } + + + &= 94 \text{ } (\alpha - \beta \text{ single crossovers}) \\
+ \beta + \text{ and } \alpha + \gamma &= 94 \text{ } (\beta - \gamma \text{ single crossovers}) \\
\alpha \beta + \text{ and } + + \gamma &= \underline{6} \text{ (double crossovers)} \\
& 1000
\end{aligned}$$

17-20. (a) $a \text{ - } b$ distance $= (2 + 1 + 69 + 76 + 145 + 143)/2000 = 21.8$ percent

$b \text{ - } c$ distance $= (5 + 5 + 2 + 1 + 69 + 76 + 1 + 1)/2000 = .8$ percent

$c \text{ - } d$ distance $= (97 + 98 + 5 + 5 + 69 + 76)/2000 = 17.5$ percent

$a \text{ - } c$ distance $= (5 + 5 + 1 + 1 + 145 + 143)/2000 = 15.0$ percent

$a \text{ - } d$ distance $= (97 + 98 + 69 + 76 + 1 + 1 + 145 + 143)/2000$
$= 31.5$ percent

$b \text{ - } d$ distance $= (97 + 98 + 1 + 1 + 1 + 1)/2000 = 10.0$ percent

The gene order is thus $a \text{ - } c \text{ - } b \text{ - } d$. On this basis, double crossovers between $a \text{ - } b$ include the classes, $a b + d, + + c +, a b + +, + + c d$, or $(5 + 5 + 1 + 1)/2000 = .6$ percent. As with all double crossovers, a frequency of twice this amount should be added to the $a \text{ - } b$ distance, which is then equal to $21.8 + 2(.6) = 23$. Double crossovers between $c \text{ - } d$ include the classes $a + c d, + b + +, a b + +, + + c d$, or $(2 + 1 + 1 + 1)/2000 = .25$ percent. Adding twice the frequency to the $c \text{ - } d$ distance gives $17.5 + 2(.25) = 18.0$. The largest distance, $a \text{ - } d$, includes the double crossovers $a + c d, + b + +, a b + d, + + c +$, or $(2 + 1 + 5 + 5)/2000 = .65$ percent, and twice this frequency (1.30) therefore should be added to the $a \text{ - } d$ distance. Note however that on the basis of the determined gene order, the classes $a b + +$ and $+ + c d$ represent *triple* crossovers. Although they have already been scored once in the *a* - *d* distance calculations as single crossovers, their frequency should now be doubled and added to the *a* - *d* distance in order to take into account the additional two crossover events that each of these classes represents [$(1 + 1)/2000 = .1$ percent $\times 2 = .2$]. Thus, the *a* - *d* distance is $31.5 + 1.3 + .2 = 33$. The linkage map is:

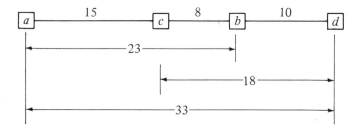

(b) *a* - *b*: Expected frequency of double crossovers = $.15 \times .08 = .012$. Observed frequency = $.006$. Coincidence = $.006/.012 = 50$ percent. Interference = 50 percent.

(c) *c* - *d*: Expected frequency of double crossovers = $.08 \times .10 = .0080$. Observed frequency = $.0025$. Coincidence = $.0025/.0080 = 31$ percent. Interference = 69 percent.

(d) Expected frequency of triple crossovers = $.15 \times .08 \times .10 = .0012$. Observed frequency = $(1 + 1)/2000 = .0010$. Coincidence = $.0010/.0012 = 83.3$ percent. Interference = 16.7 percent.

17-21. 1/22

17-22. (a) In family I, the various sons affected with sex-linked traits *a* and *b* in generation III indicate that their mother, II-2, must be carrying both sex-linked recessives *a* and *b*. Since her father's X chromosome is of genotype $a\ b^+$, and II-2 herself is normal, this means that she must have inherited an X chromosome of genotype $a^+\ b$ from her mother. That is, the X chromosome constitution of II-2 is $a\ b^+/a^+\ b$. On this basis the recombinants in family I are III-1 (*a b*) and III-5 ($a^+\ b^+$).

In family II, on the other hand, the phenotypically normal female II-1 has obviously inherited an X chromosome of constitution *a b* from her father and $a^+\ b^+$ from her mother. Thus, her offspring, III-4 ($a\ b^+$) and III-6 ($a^+\ b/a\ b$), show that recombination has occurred in her gametes. The oldest daughter of female II-1, that is, III-2, by similar reasoning, must possess the chromosomal constitution *a b* from her father (II-2) and $a^+\ b^+$ from her mother (since III-2 produces some $a^+\ b^+$ normal offspring). Thus, the recombinant offspring of III-2 are IV-2, IV-3, IV-5, IV-7, and IV-9. Note that the fact that females II-1 and III-2 are mated with males who are of genotype *a b* makes it possible to detect recombinants among both their sons and daughters. (Such daughters always inherit a paternal X chromosome, *a b*, which then enables the genotype of the maternal X chromosome to be discerned.)

(b) In family I, recombination frequency is $2/5 = 40$ percent. In family II, recombination frequency is $7/15 = 46.7$ percent.

(c) $9/20 = 45$ percent.

17-23. Since this trait is caused by a sex-linked dominant gene (e.g., D), individual I-1 in the pedigree can be carrying the dominant gene either on his X chromosome $(X^D Y^+)$ or on his Y chromosome $(X^+ Y^D)$. Regardless of whether this dominant allele began on X or Y, the daughter (II-1) must be carrying it on one of her X chromosomes $(X^D X^+)$ since she produces offspring that show the trait. Given this chromosome constitution, we can then easily follow the progress of allele D through each succeeding generation. In generation III, note that III-2 is $X^D X^+$, and the male III-3 is $X^D Y^+$ since he obtains D from his mother. In generation IV, the two male offspring of III-2 that show the trait are both $X^D Y^+$. Since III-2 is female, no X-Y crossovers of course can have occurred in her gametes. Individual III-3, however, does show the production of recombinant products since his $X^D Y^+$ chromosome constitution furnishes the recombinant gametes, X^+ (inherited by daughter IV-12), and Y^D (inherited by sons IV-9 and IV-13). By the same reasoning, individual IV-4 $(X^D Y^+)$ produces the recombinant male, V-5 (inheriting Y^D from father), and recombinant female, V-6 (inheriting X^+ from father). On the right-hand side of the pedigree, individual IV-13, who is himself a recombinant with chromosome constitution $X^+ Y^D$, now produces recombinant gametes, Y^+ and X^D, which are inherited respectively by individuals V-7 $(X^+ Y^+)$ and V-11 $(X^D X^+)$. Thus, there are a total of seven individuals who *must* be recombinants. (Note that the recombination events are only detectable in the progeny of males that show the trait, since crossovers involving the D and + alleles in this pedigree can only be discerned through the segregation of sexual phenotypes. If we were to detect recombination for this gene between the X chromosomes of females, we would need to identify segregants at an additional X chromosome locus—unavailable in the present pedigree.)

17-24. (a) The nail-patella syndrome appears to be caused by a gene inherited as a dominant (e.g., N = nail-patella syndrome, n = normal) because the trait appears in offspring of affected individuals who have mated with normals. Were it inherited as a recessive, individuals I-1 and II-1 would be heterozygotes—a notion that is inconsistent with the rarity of the trait (see text p. 109). The autosomal transmission of the trait is evident in the fact that the dominant gene causing it is not transmitted on the Y chromosome (females show the trait), nor on the X chromosome (the daughters of the two affected males, II-3 and II-6, do not show the trait; see also answers to Problems 12-22 and 12-23).

(b) The trait appears linked to the *ABO* gene in this pedigree since its appearance is generally associated with the B blood type.

(c)

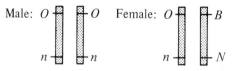

(d) II-5, II-8, II-14, III-3

(e) $4/16 = 25$ percent

(f) The male, II-3, has produced a crossover product (III-3).

(After L. S. Penrose, 1963. *Outline of Human Genetics*, 2nd ed. John Wiley, New York; from Renwick and Lawler.)

17-25. Individual III-1 presumably has the chromosomal constitution *B N/O n*. Because crossing over is not apparently restricted in human males, this individual will produce parental noncrossover gametes (*B N* and *O n*) each in frequency 1/2 $- 1/2\ r$, and recombinant gametes (*B n* and *O N*) each in frequency 1/2 *r*. Since *r* in this case is .25, the answers are as follows:

      (a) .375   (b) .125

17-26. If the genes causing type A and type B colorblindness were at the same locus, and female II-1 carried the two alleles on her X chromosomes, we would expect each of her five sons to be either type A or type B, depending on which chromosome they inherited from their mother. The presence of normal sons (III-3, III-5) indicates that crossing over has occurred between two sex-linked nonallelic genes. (Since the frequency of this event is 2/5 or 40 percent, it is much too common to be explained as the kind of intragenic recombinational event discussed in Chapter 25 and elsewhere.)

Reference: R. Vanderdonck and G. Verriest, 1960, *Biotypologie*, **21**:110–120.)

17-27. The pedigree can be diagrammed as follows:

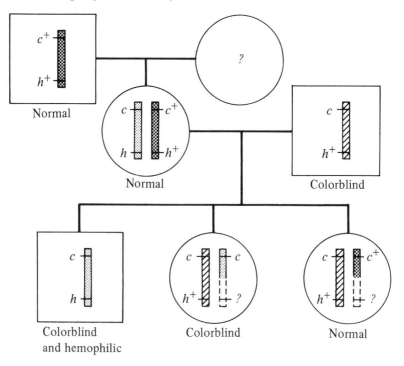

(a) The chromosome constitution of the woman II-1 can be determined by observing that she must have inherited a $c^+ h^+$ chromosome from her father who is normal for both color vision and blood clotting, as well as that she must also be carrying c and h alleles since she has produced a colorblind daughter and a hemophilic son. The question mark in the symbol for the colorblind daughter III-2 refers to the alternatives that she is heterozygous h^+/h or homozygous h^+/h^+. In the former case, she would have inherited a nonrecombinant maternal $c h$ chromosome and in the latter case a recombinant $c h^+$ chromosome. Since the expected recombination frequency between deutan colorblindness and hemophilia A is only about 7 percent as given in the problem, 93 percent of the time she would be expected to be carrying a nonrecombinant chromosome ($c h$) and therefore to be of the genotype h^+/h. Thus, the chances that she is heterozygous h/h^+ *and* that she could eventually produce hemophilic sons is .93. Given a .93 chance of being a heterozygote, the chances that any particular one of her sons would be hemophilic is $.93 \times 1/2 = .465$. (If one of her sons were born hemophilic, what are the chances the next son would be hemophilic?)

(b) The normal daughter III-3 has obviously inherited the c^+ allele from her mother, and may also have inherited the h^+ allele, which was located on the same chromosome. Since the recombination frequency between the two loci is 7 percent, there is only a .07 chance that the maternally inherited chromosome bearing c^+ also bears h as a result of crossing over. In other words, the chances for the normal daughter to be heterozygous is .07, and the chances that one of her sons will be hemophilic is $.07 \times .5 = .035$.

17-28. None.

17-29. For each of the different possible coupling relationships there is at least one son who represents a double crossover.

Possible Coupling Relationships	Double Crossover
1. $Xg^{a+} c^+ h^{B+} / Xg^{a-} c\ h^B$	II-1, II-5
2. $Xg^{a+} c^+ h^B\ /Xg^{a-} c\ h^{B+}$	II-4
3. $Xg^{a+} c\ h^B\ /Xg^{a-} c^+ h^{B+}$	II-3
4. $Xg^{a+} c\ h^{B+} /Xg^{a-} c^+ h^B$	II-2

(Reference: J.B. Graham, H.L. Tarleton, R.R. Race, and R. Sanger, 1962, *Nature*, **195**:834.)

18

Recombination in Fungi

18-1. The frequency of second-division segregation asci is equal to the frequency of meioses (tetrads) that have had an exchange between the gene in question and its centromere. (This relationship is more exact if multiple exchanges are absent in this region; since as the gene-to-centromere distance gets larger, allowing multiple exchanges, the accuracy of this map distance suffers, and the determined distance becomes less than the actual distance). Because each tetrad has two noncrossover and two crossover strands, the percentage of observed second-division segregation must be divided by 2 in order to make these distances comparable with those of organisms in which tetrad analysis cannot be performed and only single strand products of meiosis are observed (see text p. 319).

The gene-to-centromere distances are as follows:

(a) $51/(331 + 51) = 51/382 = 13.4$ percent (second division segregation); $13.4/2 = 6.7$ map units

(b) $36/(73 + 36) = 36/109 = 33.0$ percent (second division segregation); $33.0/2 = 16.5$ map units

(c) $67/(42 + 67) = 67/109 = 61.5$ percent (second division segregation); $61.5/2 = 30.8$ map units

(Reference: C.C. Lindegren, 1933, *Bull. Torrey Bot. Club*, **60**:133–154.)

18-2. (a) Because these are ordered tetrads, we can use the tetrad classification given in Table 18-1, on the basis that the *al inos* parent represents genotypes *A* and *B*, and the *+ +* parent represents *a b*. Thus the seven numbered tetrads given in the problem represent, respectively, tetrad classes ① through ⑦ of Table 18-1. On that basis, the approximate equivalence in numbers between classes ① (PD) and ② (NPD) shows no evidence for linkage. Note also that if we look at these data as unordered tetrads, tetrad classes ① and ⑤ represent PD and tetrad classes ② and ⑥ represent NPD. Again, the values are similar $(4 + 15:3 + 16)$ and show no linkage between *al* and *inos*.

    (b) Since there is no linkage between the two genes, we can follow the procedure in Table 18-1 to discern gene-centromere distances for "genes on different chromosomes." That is, we can sum the numbers of second-division segregation tetrads for each of the two genes. Thus tetrad classes ④, ⑤, ⑥, and ⑦ represent single *al* - centromere crossovers, and classes ③, ⑤, ⑥, and ⑦ represent single *inos* - centromere crossovers. The *al* - centromere distance is therefore $1/2 \times (36 + 15 + 22 + 16)/119 = 1/2 \times 89/119 = .37$. This is longer than the *inos* - centromere distance, which is $1/2 \times (23 + 15 + 22 + 16)/119 = 1/2 \times 76/119 = .32$.

18-3. Linkage appears to be absent since the NPD value (25) is certainly not less than the PD value (16). The gene-centromere distance for *sn* is zero in this experiment because there are no second-division segregations for this gene. The *cot* - centromere distance is $1/2 \times (11 + 12 + 8 + 8)/80 = 24.4$ percent. (Note that all of the tetratypes are equivalent to tetrad class ③ in Table 18-1, and represent merely single exchanges between the *B* locus in that table, or *cot* in the present example, and its centromere.)

(Reference: M.B. Mitchell, 1959, *Genetics*, **44**:847-856.)

18-4. Linkage between the two genes is obvious from the significantly lower frequency of NPD (7) compared to PD (49). The kind of linkage between the genes can be determined by noting the correspondence between tetrad classes numbered ① through ⑦ in this problem (*ad tryp* $= A B; + + = a b$) to the seven tetrad classes in Table 18-1. If we compare tetrad classes ⑤ and ⑥ according to the procedure in Table 18-1, it is obvious that the number of tetrads in the former class (8) is significantly greater ($\chi^2 = 4.00$) than in the latter (1). As explained on text p. 321, this indicates that both genes are on the same chromosome arm. The order can then be determined by noting that *tryp* shows second-division segregation in 42 tetrads, and *ad* has only 13. Thus, we can assume that the gene order is centromere - *ad* - *tryp*. On this basis *ad* is equivalent to the *A* locus in text Table 18-1 and *tryp* to the *B* locus. The exact value of the *ad* - centromere distance can then be calculated by noting that tetrad classes ④, ⑤, ⑥, and ⑦ represent single crossover tetrads for this interval, which is therefore $1/2 \times 13/100 = 6.5$ percent. Recombination in the *tryp* - centromere distance is represented by the single crossover tetrads ③ and ⑤, by the double crossover tetrads ②, ④, and ⑦, and by the triple crossover class ⑥. The *tryp* - centromere distance is therefore $1/2 (39/100) + 11/100 + 3/2 \times 1/100 = 32.0$ percent. The linkage map can then be drawn as follows:

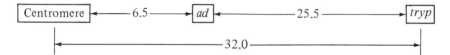

In the above case, the *ad* - *tryp* distance is calculated by subtracting 6.5 from $32.0 = 25.5$ However, this distance can also be calculated directly by noting that tetrad classes ② and ⑥, as shown in text Table 18-1, represent double crossovers for this interval, and classes ③, ④, and ⑦ represent single crossovers. The *ad* - *tryp* distance is therefore $1/2 \times 35/100 + 8/100 = 25.5$ percent.

If the data are considered entirely as unordered tetrads, then PD includes classes ① and ⑤ (57); NPD includes classes ②and ⑥ (7); and T includes classes ③, ④, and ⑦ (35). The *ad - tryp* distance is therefore (NPD + 1/2 T)/(PD + NPD + T) = [8 + 1/2 (35)]/100 = 25.5 percent. Thus, the exact same *ad - tryp* linkage distance is obtained as previously, but no centromere information is then available from these data.

18-5. There is obviously linkage between these two genes since the NPD tetrad class ② (1) has significantly fewer numbers than the PD class ① (36). Following the procedure of text Table 18-1, a comparison between classes ⑤ and ⑥ shows no difference, and we can therefore assume that the two linked genes are on opposite arms of the same chromosome. Second-division segregation for each of the two genes shows that *tryp* (classes ③, ⑤, ⑥, ⑦) is significantly farther from the centromere than *ad* (classes ④, ⑤, ⑥, ⑦). According to text Table 18-1, *tryp* can therefore be designated as locus *B* and *ad* as locus *A*. This means that tetrad class ② represents a double crossover for *tryp* but not for *ad*. The *tryp* - centromere distance is therefore 1/2 (39 + 1 + 1 + 1)/100 + 1/100 = .22, and the *ad* - centromere distance is 1/2 (21 + 1 + 1 + 1)/100 = .12. The linkage map for these genes in percentage recombination frequencies is

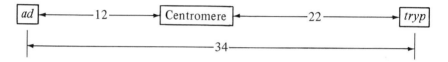

If the tetrads are considered as unordered, the centromere position cannot be obtained, but the *ad - tryp* distance can be calculated as for the previous problem:

$$(NPD + 1/2\ T)/(PD + NPD + T) = [2 + 1/2\ (61)]/100 = .325$$

Note that in this case the estimated *ad - tryp* distance differs between ordered and unordered tetrads. The reason for this is that the double crossovers between *ad* and *tryp* in classes ⑤ and ⑦ are not observed in the unordered tetrad calculations, which consider them, respectively, as PD and T.

18-6. NPD (45) is significantly less than PD (115: $\chi^2_{1df} = 30$), indicating that there is linkage between the two genes. The linkage distance based on the formula for unordered tetrads (text p. 324) is (NPD + 1/2 T)/(PD + NPD + T) = 288/646 = 45 percent.

(Reference: N.H. Giles, F.J. DeSerres, and E. Barbour, 1957, *Genetics*, **42**:608-617.)

18-7. Looked at in terms of pairs of genes, *a - c, a - b, b - c*, it is clear that the parental ditypes of the *a - c* pair $(a^+ c, a\ c^+)$ are more common than the nonparental ditypes $(a^+ c^+, a\ c)$. Specifically, there are 802 parental ditypes for these two genes, but only 2 nonparental ditypes, and 196 tetratypes. Obviously,

a is linked to *c*, and the linkage distance is $(NPD + 1/2\ T)/(PD + NPD + T)$
$= (2 + 98)/1000 = 10$ percent. Consideration of the *a* - *b* pair, on the other hand,
shows that the number of parental ditypes $(a^+ b, a\ b^+: 395 + 104 + 1 = 500)$ is
exactly equal to the number of nonparental ditypes $(a^+ b^+, a\ b: 407 + 92 + 1$
$= 500)$. Similarly, the *b* - *c* parental ditype number $(b\ c, b^+ c^+: 395 + 1 = 396)$
is actually slightly less than that of the nonparental ditypes $(b^+ c$ and $b\ c^+: 407$
$+ 1 = 408)$. The *b* gene therefore assorts independently of either *a* or *c*.

18-8. Of the 1000 tetrads given, 591 are of the parental type, indicating that
the three genes are obviously linked. Gene order can then be calculated by
various means. One method is to determine whether pairs of genes are on the
same or opposite sides of the centromere, and then to compare gene - centromere
distances. As explained on text p. 321, a comparison between tetrad classes ⑤
and ⑥ in Table 18-1 will help decide gene position with respect to the centromere.
Considering two pairs of genes at a time, the results show that classes ⑤ and ⑥
are equal for *albino* and *mating type* (1:1) and approximately equal for *albino*
and *leucine* (2:1), indicating that *albino* is on the opposite side of the centromere
to both *mating type* and *leucine*. A similar test for the frequencies of classes ⑤
and ⑥ for *mating type* and *leucine*, however, shows a definitely unequal ratio
(128:0), indicating that *leucine* and *mating type* are on the same side of the
centromere. Since more tetrads show second-division segregation for *leucine*
(266) than for *mating type* (130), this indicates that the *leucine* gene is farther
from the centromere than the *mating-type* gene, that is, the order is *albino* -
centromere - *mating type* - *leucine*.

Gene order, without centromere position, can also be determined by using
Table 18-5 in the text, and comparing the frequencies of tetrad types ①, ②,
③, and ④, as given in that table. For the present results, equating *al* + *A* to
a b c, and + *leu a* to + + +, the same sequence of tetrads in the present problem
as in Table 18-5 are tetrads ①, ⑬, ③, and ④, which have respective numbers
of 741, 1, 84, 73. Obviously the lowest observed frequency is for class ② of
Table 18-5, indicating that the present gene order is *a* - *c* - *b*, i.e., *albino* - *mating
type* - *leucine*.

When combined with the gene - centromere distance obtained from the
second-division segregation for *albino* (150), this information allows us to make
a tentative map in terms of percentage recombination frequency by multiplying
each second-division segregation frequency by one half:

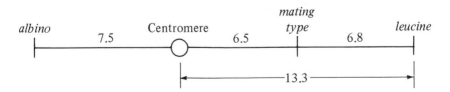

This map can be improved, however, by noting that the tetrad data includes
information on multiple crossovers as well. If we call the *albino* - centromere
distance, I, the centromere - *mating-type* distance, II, and the *mating-type* -

leucine distance, III, then the observed tetrads can be described as originating in the following manner: (The chromatids involved in the three-strand double crossovers are described in terms of the *al A* + parent providing chromatids 1 and 2, and the + *a leu* parent providing chromatids 3 and 4.)

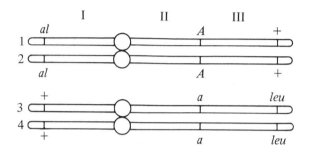

  ① Parental noncrossover
  ② Single crossover in I
  ③ Single crossover in III
  ④ Single crossover in II
  ⑤ Two-strand double crossover in I and II
  ⑥ Three-strand double crossover (chromatids 2–3 and 1–3) in I and II
  ⑦ Three-strand double crossover (chromatids 2–4 and 2–3) in I and II
  ⑧ Four-strand double crossover in I and II
  ⑨ Two-strand double crossover in I and III
  ⑩ Three-strand double crossover (chromatids 2–3 and 1–3) in I and III
  ⑪ Three-strand double crossover (chromatids 2–3 and 2–4) in I and III
  ⑫ Three-strand double crossover (chromatids 2–3 and 2–4) in II and III
  ⑬ Four-strand double crossover in II and III
  ⑭ Triple crossover (chromatids 1–4, 2–3, and 1–3) in I, II, and III

Exact linkage distances for each of the three regions, I, II and III, can then be determined by using the listed information to score the frequencies of crossover tetrads in each area and multiplying by 1/2 (text pp. 325–327). Region I = 1/2 (150/1000) = 7.5 percent; region II = 1/2 (130/1000) = 6.5 percent; region III = 1/2 (140/1000) = 7.0 percent.

18-9. If we call the *y* - *sn* distance I, and the *sn* - centromere distance II, the two somatic X chromosomes can be visualized as follows:

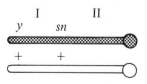

A crossover in region I can then produce a *yellow* and wild-type pair of cells (arrows indicate the direction of cell-division segregation):

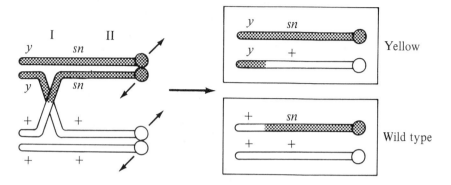

A crossover in region II can produce a *yellow singed* and wild-type pair of cells:

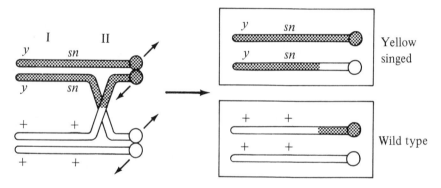

A double crossover may produce a *singed* and wild-type pair of cells:

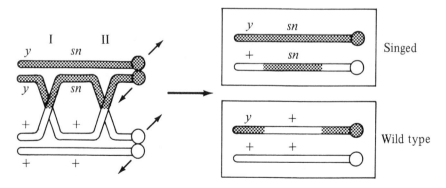

Because a wild-type cell is produced as a result of each mitotic crossover event, there will be no appearance of twin spots bearing *yellow* and *singed* tissues side-by-side, and it might be difficult to ascribe the appearance of mutant tissue to somatic (mitotic) recombination rather than to somatic mutation.

18-10. The most likely explanation is that *ribo* 1 and *ad* 14 are on one arm of the chromosome, whereas *pro* 1 and *bi* 1 are on the other arm. The relative positions would be *ribo* 1 - *ad* 14 - *centromere* - *pro* 1 - *bi* 1. Although they are all linked, *ribo* 1 and *ad* 14 appear to assort independently from *pro* 1 and *bi* 1, indicating enough distance between the two pairs of genes that the crossover frequencies are close to 50 percent or higher. This is much more likely if *ribo* 1 and *ad* 14 are on one arm and *pro* 1 and *bi* 1 on the other arm.

18-11. In strain Y, all *w* homozygotes are also homozygous *Acr*, whereas not all *Acr* homozygotes are homozygous for white. This indicates that the sequence is centromere - *w* - *Acr*, and the diploid chromosomes are as follows:

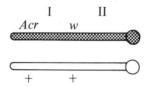

A crossover in region II followed by the proper segregation of centromeres will give cells that are *w Acr/ w Acr* and w^+Acr^+/w^+Acr^+. All cells that are *w/w* are also *Acr/Acr*:

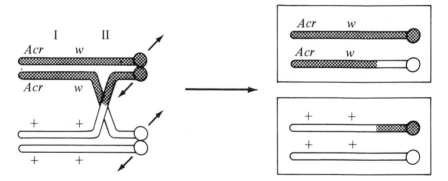

If, however, we look at the *Acr/Acr* individuals, those 87 percent that are also *w/w* arose from an exchange in region II (see above), whereas those 13 percent that are w/w^+ arose from an exchange in region I:

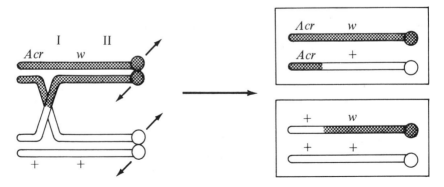

This indicates that the centromere - w distance is much greater than the w - Acr distance.

The data from the Z strain, in which the two genes are in repulsion, are consistent with those from the Y strain, with only slight differences in the frequency of crossovers in the two regions.

The absence of detectable double crossovers (white, nonacriflavine-requiring in strain Y; and white, acriflavine-requiring in strain Z) could indicate that the overall crossover map distances (centromere - w - Acr) are not large enough for an easily detectable frequency of double crossovers to occur.

18–12. The w-3 pu ad-1 loci act as a unit and appear to assort independently of y (y and w-3 pu ad-1 = 18; y and + + + = 23). Similarly, the sm $phen$-2 loci act as a unit, and appear to assort independently of y (y and sm $phen$-2 = 23; y and + + = 18). The w-3 pu ad-1 loci appear to be assorting independently of the sm $phen$-2 loci, with the four combinations in a ratio of 7:11:16:7 (x^2_{3df} = 5.36; probability = .2–.1). There appear to be three separate linkage groups:

1. y
2. w-3 pu ad-1
3. sm $phen$-2

19

Recombination in Bacteria

19-1. (a) Isolate and highly purify the DNA from the wild-type strain, and mix it with cells of the mutant strain. (Keep each strain in a separate culture to prevent contact between mutant cells and intact wild-type cells, thereby ruling out conjugation.)

(b) Use cell-to-cell contact between the two strains in the presence of DNase. The DNase would eliminate the possibility of transformation, since it would digest any transforming DNA.

(c) The mutant and wild-type strains should both be lysogenic for the same temperate phage, and cell-to-cell contact between the two strains should be prevented by keeping each strain in separate cultures. Obtain lytic products (transducing phage) from the wild-type strain, and subject this material to DNase. This will destroy any transforming DNA that might be present, but will not affect the mutant bacteria that are to be transduced.

19-2. Only 1 of every 200 DNA molecules would carry this gene. If each bacterium can absorb 8 molecules, then the maximum frequency of expected transformants is $1/200 \times 8 = 8/200 = 4$ percent.

19-3. On the assumption that the *his* gene is located at one end, the transformation linkage map is

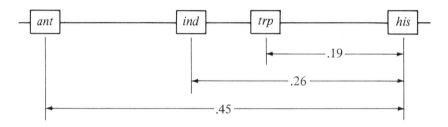

(Reference: C. Anagnostopoulos and I.P. Crawford, 1961, *Proc. Nat. Acad. Sci.*, 47:378-390.)

19-4. As shown on text p. 344, the linkage distance between two genes is calculated as the number of recombinants between the two genes divided by the total number of transformants for these genes. The *pro* - *ala* distance in this problem, for example, is calculated as the number of $pro^+ ala^-$ and $pro^- ala^+$ divided by the number of individuals who are either pro^+, or ala^+, or both $pro^+ ala^+$. The calculations are therefore as follows:

(a) *pro* - *ala* = (200 + 500 + 100 + 400)/ (7,200 + 200 + 500 + 800 + 100 + 400) = 1200/9200 = 13 percent.

pro - *arg* = (500 + 800 + 400 + 800)/ (7,200 + 500 + 800 + 100 + 400 + 800) = 2500/9800 = 26 percent.

ala - *arg* = (200 + 800 + 100 + 800)/ (7,200 + 200 + 500 + 800 + 100 + 800) = 1900/9600 = 20 percent.

(b) Based on the observed transformation linkage distances, the linkage order appears to be *pro* - *ala* - *arg*:

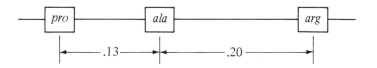

The same linkage order can be obtained by noting that the least frequent class ($pro^+ ala^- arg^+$) arises from a quadruple crossover if the *pro* - *ala* - *arg* gene order is correct, whereas the other more common classes are all the result of only double crossovers:

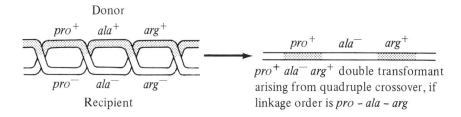

19-5. (a) The most frequent transformant classes would be those that show transformation at the linked loci with absence of transformation at *c*, that is, $a^+ b^+ c^-$, or transformation at *c* with absence of transformation at *a* and *b*, that is, $a^- b^- c^+$. The least frequent classes would be those that show simultaneous transformation for the two unlinked segments, $a^+ b^+ c^+$, or transformation for both segments with a crossover in one, for example, $a^+ b^- c^+$.

(b) The most frequent would be $(a^+ c^+) b^-$ or $(a^- c^-) b^+$, whereas the least frequent would again be $a^+ b^+ c^+$.

(c) The least frequent would be that class which results from a quadruple crossover:

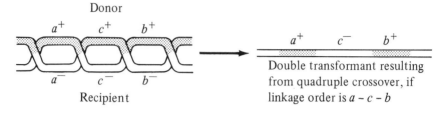

Donor

Double transformant resulting from quadruple crossover, if linkage order is $a - c - b$

Recipient

(d) As in (c), but now $c^+ a^- b^+$

19-6. Using the test for independence (text Chapter 8, pp. 133–134), χ^2 for 1 degree of freedom is calculated as 29.4. This indicates very strongly that phage T1 resistance does not assort independently of the other traits, and that it is linked to them.

(Reference: J. Lederberg, 1947, *Genetics*, **32**:505–525.)

19-7. (a)

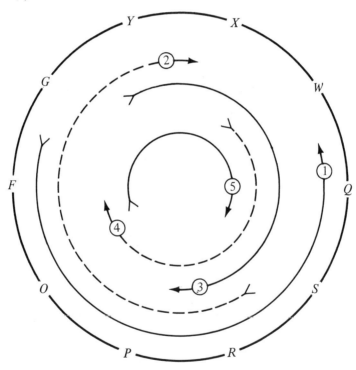

(b) $1 = W$; $2 = X$; $3 = P$; $4 = F$; $5 = S$

19-8. The largest number of prototrophs among the surviving recipients are of the c^+ type, indicating that the donor c^+ gene had the greatest opportunity for recombination, and must therefore have entered first among the four genes scored.

On the other hand, the absence of b^+ prototrophs indicates that the donor b^+ gene must be located at that "end" of the donor chromosome that does not have the opportunity for entry and recombination. By similar reasoning the a^+ gene enters after c^+ but before d^+, and the donor sequence can be mapped as follows:

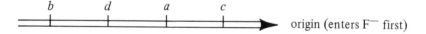

origin (enters F^- first)

19-9. At the distal end, so that the streptomycin sensitivity remains in the donor.

19-10. The order is *cysB - trpB - trpD* since the relatively high frequency of *cysB trp*$^+$ products in the first cross can be explained as arising from a double crossover, whereas the relatively low frequency of such products from the second cross arises from a quadruple crossover according to this gene order. If the gene order were *cysB - trpD - trpB*, the *cysB trp*$^+$ products would have arisen from a quadruple crossover in the first cross and from a double crossover in the second. This would mean that quadruple crossovers are more common than double crossovers—a notion that is inconsistent with all other findings.

(Reference: M. Demerec and Z. Demerec, 1956, *Brookhaven Symp. Biol,* 8:75-87.)

19-11. As expected, no prototrophs appear when donor and recipient are both $B1^-$ or $B3^-$. The appearance of prototrophs among recipients occurs only when donor strains can supply a + locus for a − locus on the recipient. However, for a + donor locus to be integrated into a recipient chromosome, a crossover must occur, and the frequency of such crossing over indicates the distance between the three *B* loci. For example, if we look only at those loci for which the donors and recipients differ, then the prototroph frequencies of (c) and (d) in the table given for this problem arise from the following events:

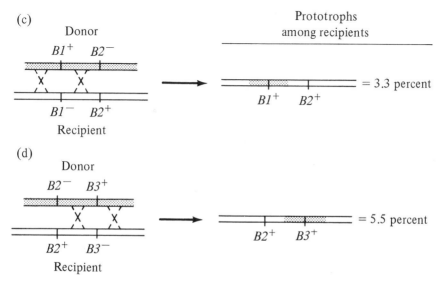

These findings therefore indicate that crossovers occur more frequently between *B2* and *B3* (5.5 percent) than between *B1* and *B2* (3.3 percent), and the distance between the former is greater than between the latter. By the same reasoning, (b) and (e) indicate that the *B1 - B3* distance is either 1.7 percent or 1.8 percent. Thus, a map of the three loci can be drawn as follows:

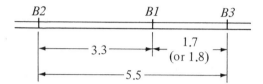

(Reference: C. Yanofsky and E.S. Lennox, 1959, *Virology*, 8:425–447.)

19-12. The *recB* product is probably responsible for a considerable portion of the observed degradation, since its presence without the *recA* product (*rec A⁻ B⁺*) increases degradation, whereas its absence (*rec A⁺ B⁻* or *rec A⁻ B⁻*) reduces degradation.

(Reference: A.J. Clark, 1973, *Ann, Rev. Genet.*, 7:67–86.)

20

Recombination in Viruses

20-1. Because of phenotypic mixing, (see text p. 377) phages are produced with c^+ protein envelopes but c genotypes.

20-2. (a) There are two sites on the viral capsule involved in host range specificity, each site capable of being formed independently of the other through random association of the proteins produced by the different viruses that infect a cell. As a result of phenotypic mixing, one of these sites in some viruses is now associated with T2 protein (thereby capable of infecting B/4), and the other site is associated with T4 protein (capable of infecting B/2). The DNA, of course, is either T2 or T4 and therefore produces pure strains of viruses on further growth.

     (b) Assuming that each strain of virus produces an equal amount of protein for each of these two sites, which may then associate randomly: T2 X T4 → 1 T2/T2:1 T4/T4:2 T2/T4. In other words, about half the products should have mixed phenotypes.

(Reference: G. Streisinger, 1956, *Virology*, 2:388-398.)

20-3. The T2 X T2 hybridization should yield more heavily labeled DNA strands since T2 chromosomes are circularly permuted (see text p. 385), and therefore provide many different kinds of "ends" that can hybridize with various sections of the chromosome:

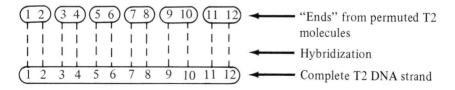

"Ends" from permuted T2 molecules

Hybridization

Complete T2 DNA strand

Phage T5, on the other hand, is not circularly permuted (text p. 385), and there will be only two kinds of "ends" at most for hybridization:

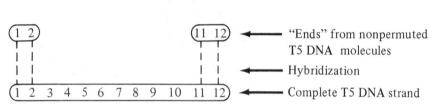

"Ends" from nonpermuted T5 DNA molecules

Hybridization

Complete T5 DNA strand

(Reference: C.A. Thomas, 1963, *Cold Sp. Harb. Symp.*, 28:395-396.)

20-4. As explained in the text, the formation of circles depends on redundancy at both terminals of a chain of nucleotides. Fragmentation of such DNA, however, leads to nucleotide chains that are no longer terminally redundant at both ends, and circles would not be expected.

20-5. An increase in heterozygote frequency would be expected since a T2 "headful" (text p. 387) has a greater probability of carrying two *h* loci when an intervening chromosome section is deleted:

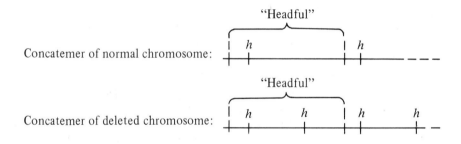

20-6. (a) $(311 + 341)/12314 = 5.3$ percent
(b) $(65 + 56)/1270 = 9.5$ percent
(c) $(19 + 20)/1413 = 2.8$ percent
(d) $(84 + 75)/2577 = 6.2$ percent

Since only one cross with co_1 is given, the gene can be either to the right or left of *mi*. The following linkage map is therefore arbitrary in respect to the position of this locus.

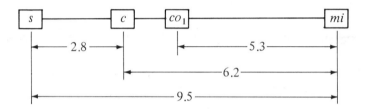

20-7. (a) m - $r = (162 + 520 + 474 + 172)/10342 = 12.84$ percent
r - $tu = (853 + 162 + 172 + 965)/10342 = 20.81$ percent
m - $tu = (853 + 520 + 474 + 965)/10342 = 27.19$ percent

(b) m - r - tu. The distance between m - tu therefore is increased by twice the double crossover frequency $[m + tu, + r +, = (162 + 172)/10342 = 3.23$ percent] and is therefore equal to $27.19 + 2(3.23) = 33.65$.

(c) Coincidence $= \dfrac{.0323}{(.1284 \times .2081)} = \dfrac{.0323}{.0267} = 1.21$

This signifies the absence of interference in this cross or perhaps even the presence of negative interference.

(Reference: A.H. Doermann, 1953, *Cold Sp. Harb. Symp.*, **18**:3-11.)

20-8. Assuming that the proximity of DNA strands is related to the possibility of recombination, a localized DNA pool (T2) would provide more opportunities for recombination between DNA molecules than one that is more diffuse (T5) or one that is so small or diffuse that it cannot be seen (λ). These observations would therefore help explain the finding that recombination frequencies are in the order $T2 > T5 > \lambda$.

20-9. In T4 the concatemers are cut into chromosome lengths at different places (circular permutations) so that there are no "end" genes. That is, genes that are separated in some T4 chromosomes are linked in others. In λ, on the other hand, the cuts that transform the vegetative circle into a linear chromosome are always at the same specific point (m-m'; see text Fig. 19-18). Thus, the λ genes on each side of the cut are always separated and do not appear to be linked (except in the prophage where m and m' occupy adjacent positions; see, for example, Figure 19-19d).

20-10. See the accompanying figure on the following page. (a) 33.3 (relative length of labeled DNA transferred to recombinant from I^+ strain) $\times$ 1/2 (proportion of labeled DNA that remains after a single replication in unlabeled medium) $= 16.7$ percent.

(b) No labeling.

(c) Answer to (a) would be $.66 \times 1/2 = 33.3$ percent, and the answer to (b) would again be no labeling.

Figure for Problem 20–10

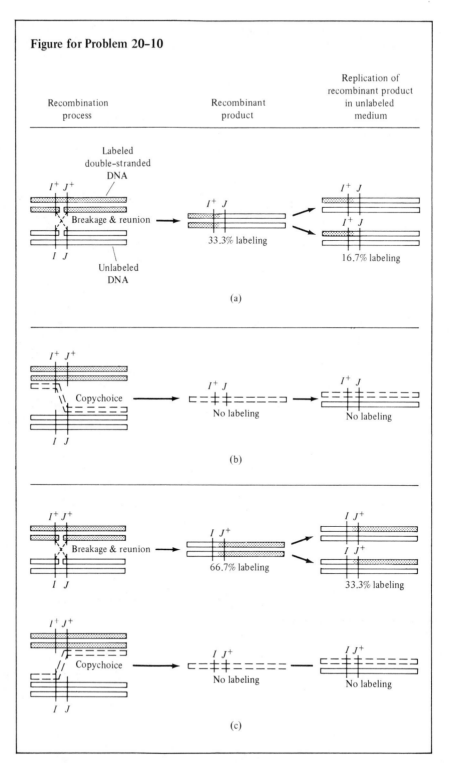

Recombination process — Recombinant product — Replication of recombinant product in unlabeled medium

(a)

(b)

(c)

21

Chromosome Variation in Number

21-1. The triploid will contain $12 + 6 = 18$ chromosomes (3n) in the form of 6 linkage groups (n = 6). Assuming that each gamete gets one complete set (6) through the normal disjunction of 6 pairs of chromosomes, and that the 6 unpaired chromosomes segregate randomly to opposite poles (3 to each), then the expected average number of chromosomes per gamete is $6 + 3 = 9$. Since an average of only 7 are found, there is an average loss of 2 chromosomes.

21-2. (a) Allopolyploidy. The sterile diploid hybrid produced tetraploid tissue, which could then, by homologous pairing, produce normal meiotic products.

(b) The allopolyploid contains a diploid *verticillata* and a diploid *floribunda* set of chromosomes. Since each of these parental species contributes a haploid number of 9 chromosomes, the allopolyploid can form 9 bivalents for each parental set of chromosomes, or a total of 18 bivalents:

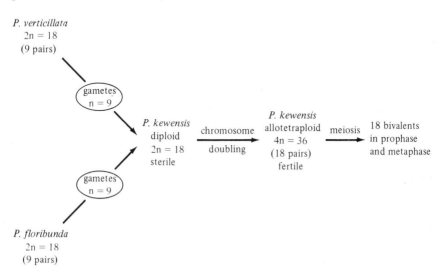

(Reference: W.C.F. Newton and C. Pellew, 1929, *J. Genet.*, **20**:405–467.)

21-3. The "rabbage" is a fertile allopolyploid of 36 chromosomes, or 18 pairs.

Since 9 of these pairs derive from the 9-chromosome gamete of the radish parent, the other 9 pairs of chromosomes must derive from a 9-chromosome gamete contributed by the cabbage parent. (See the accompanying figure, and also previous Problem 21-2.)

21-4. Assuming there are no crossovers between the A locus and its centromere, the answers are as follows:

    (a) $1AA:1Aa$ (b) $1Aa:1aa$

21-5. (a) With AA and Aa gametes, the tetraploid progeny will be: $1/4\ AAAA$: $2/4\ AAAa:1/4\ Aaaa$, and all will have the A phenotype.

    (b) With Aa and aa gametes, the tetraploid progeny will be: $1/4\ AAaa$: $2/4\ Aaaa:1/4\ aaaa$. The phenotypic ratio will be 3/4 dominant (A - - -) and 1/4 recessive ($aaaa$).

21-6. Assuming random chromosome segregation in the 4n autotetraploid, 2n gametes for the A locus are produced in the ratio of $1/6\ AA:4/6\ Aa:1/6\ aa$, or $5/6\ A$-:$1/6\ aa$. Upon self-fertilization, the progeny that result are therefore $1/6\ aa \times 1/6\ aa = 1/36\ aaaa$, and all the rest (35/36) are A - - -. Similarly, random chromosome segregation at the B locus yields progeny in the ratio $35/36\ B$ - - -: $1/36\ bbbb$. The answer is therefore $(35/36)^2$ AB:$(35/36)(1/36)$ Ab:$(1/36)(35/36)$ aB:$(1/36)^2$ ab phenotypes.

21-7. (a)

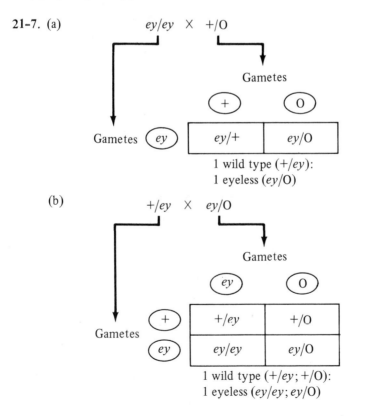

$ey/ey \times +/O$

Gametes

1 wild type ($+/ey$):
1 eyeless (ey/O)

(b)

$+/ey \times ey/O$

Gametes

1 wild type ($+/ey$; $+/O$):
1 eyeless (ey/ey; ey/O)

Figure for Problem 21-3

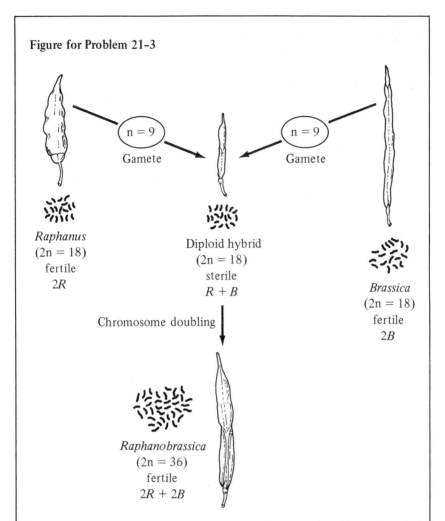

Seed pods and chromosomes of the radish (*Raphanus*), cabbage (*Brassica*), and their allotetraploid hybrid (*Raphanobrassica*) for Problem 21-3. The diploid hybrid is formed by combining a gamete of *Raphanus* (n = 9) with a gamete of *Brassica* (n = 9) and has a chromosome constitution of 2n = R + B = 18. Because most radish and cabbage chromosomes fail to pair interspecifically, meiosis in the diploid hybrid is abnormal, making the plant almost entirely sterile. Only when the chromosomes of this diploid double to form the illustrated allotetraploid is a plant with full fertility attained, since the 9 radish and 9 cabbage chromosomes now have their own specific pairing mates.

(From G.D. Karpechenko, 1928, Polyploid hybrids of *Raphinus sativas* L. X *Brassica oleracea* L. *Zeit. Induk. Abst. u. Vererbung.*, 48:1-85.)

21-8. The answers below are based on the following assumptions: the trisomic produces gametes with random constitutions of one or two fourth chromosomes; the monosomic produces equal numbers of monosomic and nullisomic gametes; the presence of the wild-type allele for *bent* produces the wild-type phenotype even when there is more than one *bent* mutant allele; and zygotes without a fourth chromosome, or with four fourth chromosomes, do not reach the adult stage.

(a) 2 wild type ($+ /bt/bt$; $+ /bt$):2 bent (bt/bt; bt)

(b) $+ /bt/bt \times + /bt/bt \to$ 19 wild type:8 bent $+ /bt/bt \times + /bt \to$ 3 wild type:1 bent

$+ /bt/bt \times bt/bt \to$ 1 wild type:1 bent $+ /bt/bt \times bt \to$ 1 wild type:1 bent

$+ /bt \times + /bt \to$ 3 wild type:1 bent $+ /bt \times bt/bt \to$ 1 wild type:1 bent

$+ /bt \times bt \to$ 1 wild type:1 bent $bt/bt \times bt/bt \to$ all bent

$bt/bt \times bt \to$ all bent $bt \times bt \to$ all bent

21-9. (a) $.50 \times .10 = .05 = 5$ percent
(b) $(.50 \times .90) + (.50 \times .10) = .50 = 50$ percent
(c) $.50 \times .90 = .45 = 45$ percent

21-10. (a) I: $A \cdot B, A \cdot B, A \cdot A$. II: $A \cdot B, B \cdot A, A \cdot C, C \cdot D, D \cdot C$, or $A \cdot B, B \cdot C$, $C \cdot D, D \cdot E, E \cdot F$

(b) I: secondary trisomic; II: tertiary trisomic (text p. 412), or multiple translocation (text Chap. 22)

21-11. (a) X:A ratio $= 2/4 =$ male
(b)   '' $= 3/4 =$ intersex
(c)   '' $= 0/4 =$ super-super male (lethal)
(d)   '' $= 2/4 =$ male
(e)   '' $= 4/4 =$ female

21-12. (a) If we designate C^+ as the normal allele at the potato-leaf locus, the progeny in the problem derive from the crosses shown at the top of the following page. If the normal allele C^+ is dominant over c in any dosage, the expected results are therefore 5 normal-leaved ($C^+ -$):1 potato leaf (cc).

(b) This is a simple cross between disomics, $C^+C^+ \times cc$. Thus, the C^+c F_1, when backcrossed to cc, will produce progeny in the ratio of 1 normal leaf:1 potato leaf.

21-13. (a) The monosomic plants will be white. The nonmonosomic (disomic) plants will be red.

(b) All the progeny will be red.

Figure for Problem 21-12

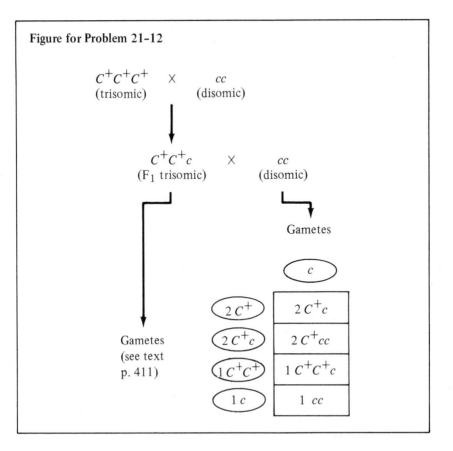

$c^+c^+c^+$ × cc
(trisomic) (disomic)

c^+c^+c × cc
(F_1 trisomic) (disomic)

Gametes

Gametes
(see text
p. 411)

	c
$2\,c^+$	$2\,c^+c$
$2\,c^+c$	$2\,c^+cc$
$1\,c^+c^+$	$1\,c^+c^+c$
$1\,c$	$1\,cc$

21-14. The F_1 trisomics would be $su/su/Su$ and would produce the following gametophytic ratios: .33 su :.17 su/su :.17 Su :.33 Su/su, or one half carrying the dominant Su allele and one half with only the recessive allele(s). When self-fertilized, they would produce an F_2 of disomics, trisomics, and tetrasomics, with a phenotypic ratio of 3 Su:1 su. Since this is also the normally expected ratio for diploid segregation, the data do not permit us to determine on which chromosome Su is located.

21-15. (a) In offspring with 35 chromosomes, the *N. tabacum* plant has contributed 23 and the *N. sylvestris* plant has contributed 12. If the monosomic *N. tabacum* parent had been lacking a *tomentosa* chromosome, pairing should have occurred with the 12 *sylvestris* homologues, and there should be 12 bivalents and 11 univalents. However, since there are 11 bivalents and 13 univalents, this indicates that a *sylvestris* chromosome is unpaired, and this unpaired chromosome is now added to the 12 *tomentosa* univalents.

(b) 12 bivalents and 11 univalents, as explained above.

21-16. (a) ABB × $aabb$ gives one half $AaBb$ (green disomic) and one half aBb (green monosomic).

(b) The cross between F_1 monosomics $aBb \times aBb$ gives three fourths $B-$ (green) and one fourth bb (yellow), since the A locus carrying only a alleles does not interfere with color determination by the B locus.

$AaBb \times AaBb$ is a standard dihybrid cross, where 1/16 of the progeny will be $aabb$ (yellow) and the remaining 15/16, with at least one dominant allele, will be green.

21-17. Since there are two pairs of genes involved, we can describe the *yellow burley* variety as *aabb* and the green plants as *AABB*. If neither the *A* nor the *B* genes are on monosomic chromosomes, then the F_1 is *AaBb*, and the expected F_2 ratio (*AaBb* $\times$ *aabb*) is 3 green:1 yellow burley. However, if the *A* or *B* genes are located on one of the monosomic chromosomes, then the F_1 can be described as *aBb*, and when crossed with yellow burley (*aBb* $\times$ *aabb*) produces the ratio 1 green (*aaBb*, *OaBb*):1 yellow burley (*aabb*, *Oabb*). Since the latter ratio is observed only for the *O* monosomics, one of the *yellow burley* genes must be located on that chromosome.

21-18. If a susceptible gene is not on a trisomic chromosome, the F_1 cross can be described as $Ss \times ss$ and should yield a ratio of 1 susceptible: 1 resistant. If the gene involved is located on a trisomic chromosome, the F_1 cross is $Sss \times ss$, and other than normal ratios will result (see text Table 21-5). In the present case, the gene is probably located on chromosome *10* since the trisomic for this chromosome produces a ratio of approximately 2 susceptible:1 resistant. It is likely that this abnormal ratio is caused by the inviability of aneuploid gameto- phytes (see Problem 21-9), a majority of which would be carrying the S allele, whereas a majority of the euploid gametophytes would be s.

(Reference: V.H. Rhoades, 1935, *Proc. Nat. Acad. Sci.*, **21**, 243–246.)

21-19. In bees, the male parts are haploid and the female parts are diploid. Among the possible explanations that can be offered are therefore the following:
    (a) Double fertilization of one egg occurred with two sperm. One sperm nucleus fertilized the egg nucleus to produce the diploid female parts, whereas the other sperm nucleus replicated independently to produce the haploid male parts, which are therefore of paternal origin:

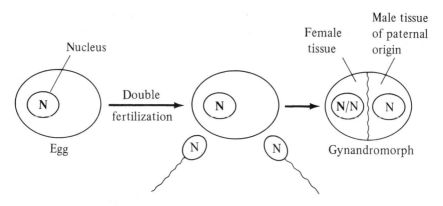

(b) An egg containing two maternal nuclei was fertilized by a single sperm. Fusion between the sperm nucleus and one of the egg nuclei occurred to produce the diploid female parts, while independent replication of the other nucleus occurred to produce the haploid male parts, which are therefore of maternal origin:

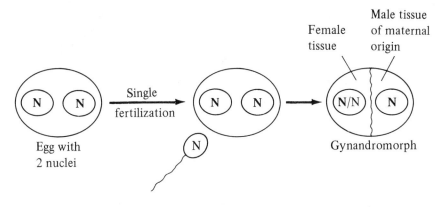

(For other mechanisms producing gynandromorphs in Hymenoptera, see K.W. Cooper, 1959, *Bull. Florida State Mus. Bio. Sci.*, 5:25-40.)

21-20. On the basis that at least one 21-chromosome goes to each pole of the first meiotic division, the trisomic individual can produce gametes that contain either two 21s or one 21. Assuming that tetrasomics for chromosome 21 are not viable, the ratio among surviving offspring will therefore be 2 affected:1 normal:

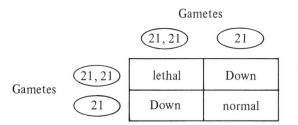

21-21. (a) Since the woman claims that she has not been fertilized by a Y-bearing male, her child can only be either XX (normal female) or XO (female with Turner's syndrome).

(b) If the zygote arose through parthenogenesis in which the maternal diploid complement is preserved as the diploid complement of the egg nucleus, then the mother and child should be identical in phenotype for all genetic traits, such as blood groups.

21-22. We assume the Klinefelter male is fertile (they are, in fact, sterile). We also assume that there is an equal chance for meiotic pairing to occur between any two of his sex chromosomes, and that his gametes contain either one or two of the three sex chromosomes. Thus, the XXY male may produce three kinds of equally possible pairing combinations: $X^1 X^2 + Y; X^1 Y + X^2; X^2 Y + X^1$. Since each Y-containing gamete produces a son, two out of three sons are expected to show Klinefelter's syndrome.

21-23. If there are equal chances for meiotic pairing between any two of the three sex chromosomes in an XYY male, and all resulting meiotic products have equal chances of being incorporated into gametes, then the gametes of such a male are expected in the ratio 2 male Klinefelter's (XXY) : 1 male trisomic (XYY) : 2 normal male (XY) : 1 normal female (XX). (In fact, however, most of the progeny of XYY males are normal XX females and XY males, for reasons that are not yet clear.)

21-24. The paternal X chromosome must be missing since, if it were present, the female with Turner's syndrome would have normal color vision.

21-25. Since both parents are phenotypically normal, and red-green color-blindness is sex-linked recessive, the mother must be the heterozygous carrier. The colorblind Klinefelter's child (XXY) must therefore have inherited two maternal X chromosomes, both carrying the colorblind gene. That is, the nondisjunctional event must have taken place in the second meiotic division of maternal gametogenesis since it is at this stage that two identical chromosomes can accidentally segregate to the same pole (see text Fig. 12-6).

21-26. (a)

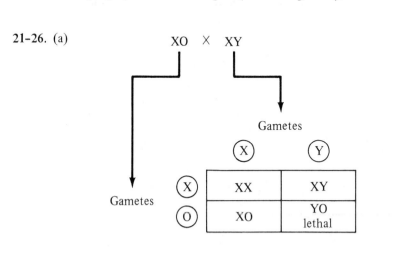

The ratios are therefore 1 XX (female) : 1 XY (male) : 1 XO (female) : 1 YO (dies).

(b) Assuming that there are equal chances for meiotic pairing between any two of the three sex chromosomes in the XXY male, with the unpaired chromosome going at random to either pole, the cross can be pictured as follows:

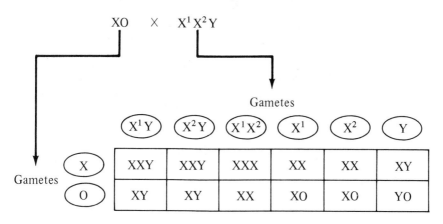

The ratios are therefore 3 XX (female) : 3 XY (male) : 2 XO (female) : 2 XXY (male) : 1 XXX (superfemale?) : 1 YO (dies).

21-27. Nondisjunction in a "calico" female can produce an $X^Y X^y$ egg that is fertilized by a Y-bearing sperm, yielding a trisomic "Klinefelter's" male of genotype $X^Y X^y Y$. Obviously this nondisjunctional event would have occurred at the first meiotic division (see text Fig. 12-6).

21-28. Note that the calico male is, by definition, XXY, carrying *yellow* on one X chromosome and *black* on the other. The only parent that could have provided the *yellow*-bearing X chromosome in this pedigree is the father (II-2) who also provided the Y chromosome. Thus, the nondisjunctional event must have occurred in the first paternal meiotic division (see text Fig. 12-5).

21-29. A simple explanation is provided by the Lyon hypothesis (text pp. 421-422) that it is a random matter which of the two X chromosomes in a mammalian female is inactivated during development. In the present example, we know that the paternal X chromosome carries a colorblind allele and can assume that the maternal X chromosome contributed to the twins carries the normal allele. Thus, it is presumably the maternal X that is most frequently inactivated in the optical pathways of the colorblind twin, whereas the normal-sighted twin has its paternal X most frequently inactivated.

(Reference: J. Philip et al., 1969, *Ann. Hum. Genet.*, **33**:185-195.)

22

Changes in Chromosome Structure

22-1. The usual mutation rate for a simple gene is of the order of 1 per 100,000, whereas the present data indicated a mutation rate that is more than 2 orders of magnitude greater (400 per 100,000). One explanation is therefore that the x-ray treatment has caused a chromosomal deficiency and the appearance of the recessive phenotype in the F_1 is caused by pseudodominance (text p. 430). Cytological attempts should be made to look for the chromosomal deletion, or genetic attempts should be made to look for a reduction in recombination frequencies between genes lying on each side of the deficiency.

22-2. The transmission of C-carrying chromosomes through the pollen can be explained as arising from crossovers that enable the deleted section on the C chromosome to be replaced by a complete section from the normal chromosome:

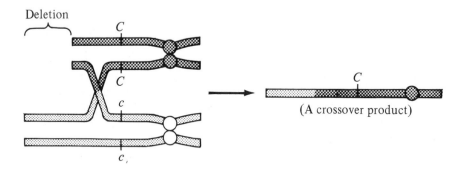

(A crossover product)

22-3. Since *prune* is a sex-linked recessive gene its appearance in the F_1 females would not be expected unless the *Notch* deficiency included the loss of the *prune* locus. The appearance of *Notch* F_1 females that are prune-eyed means that these females are hemizygous (pseudodominant) for *prune*, and the *Notch* deficiency therefore includes the *prune* locus. The subsequent cross is therefore *Notch/prune* ♀ X *prune*$^+$ ♂ → 1 notch-winged female (*Notch/prune*$^+$): 1 wild-type female (*prune/prune*$^+$): 1 prune-eyed male (*prune*). The *Notch* class of males is absent because of lethality.

22–4. Pairing in the inversion heterozygote would look like the following (For diagrammatic simplicity, only the paired chromosomes without chromatids are shown.):

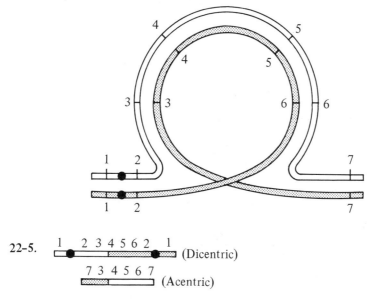

22–5.

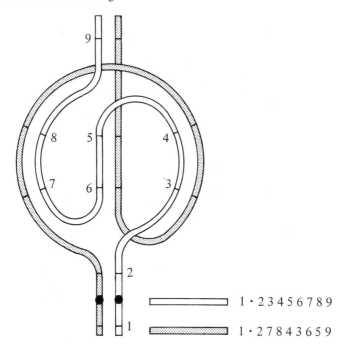

22–6. Without showing individual chromatids, pairing of the two chromosomes would look like the following:

Note that the double loop indicates that two inversions have occurred, one *overlapping* the other:

$$1 \cdot 2\ 3\ 4\ \boxed{5\ 6\ 7\ 8}\ 9$$

$$1 \cdot 2\ \boxed{3\ 4\ \boxed{8\ 7}\ 6\ 5}\ 9$$

$$1 \cdot 2\ \boxed{7\ 8\ 4\ 3}\ 6\ 5\ 9$$

(Each box that includes gene numbers in two adjacent chromosomes indicates the inversion that has taken place between them.)

22-7. (a)

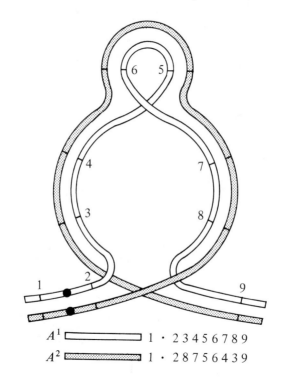

$A^1 \ \rule{3cm}{0.4cm}\ 1 \cdot 2\ 3\ 4\ 5\ 6\ 7\ 8\ 9$

$A^2 \ \rule{3cm}{0.4cm}\ 1 \cdot 2\ 8\ 7\ 5\ 6\ 4\ 3\ 9$

Note that the presence of one loop (5–6) within another (3–8) indicates that a double inversion has occurred, one *included* within the other.

(b)

$$1 \cdot 2\ \boxed{3\ 4\ 5\ 6\ 7\ 8}\ 9 \quad \text{Original sequence } (A^1)$$

$$1 \cdot 2\ 8\ 7\ \boxed{6\ 5\ 4\ 3}\ 9 \quad \text{Intermediate sequence}$$

$$1 \cdot 2\ 8\ 7\ \boxed{5\ 6}\ 4\ 3\ 9 \quad \text{Final sequence } (A^2)$$

(Each box spanning two adjacent gene sequences indicates an inversion between them.)

22-8. The intervals in which recombination frequencies are most greatly reduced are *vg* - *L* and *L* - *px*. We can assume, therefore, that the inversion covers these two sections, that is, the two chromosomes look as follows (Numbers are used only to indicate unnamed gene sequences for illustration purposes.):

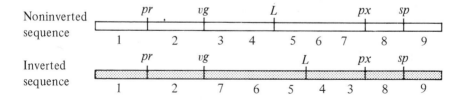

Pairing in the inversion heterozygote (individual chromatids omitted) is therefore as follows:

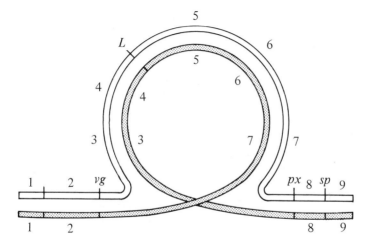

22-9. Arrangement 3 arises by two breaks from either arrangement 1 or arrangement 2. The difference between arrangements 1 and 2, however, is no less than four breaks. Thus, the sequence is probably 1 ↔ 3 ↔ 2.

(1) A B C|D E F G H|I J

(3) A B C|H G F|E D|I J

(2) A B C H G F|I D E|J

(Inversions between adjacent sequences are indicated by a box.)

22-10. If we indicate inversion breakage points by arrows (↓), then the original arrangement *AB↓CDEFGH↓IJ* produced arrangement (a) by the indicated inversion. Arrangement (b) was produced by two nonoverlapping inversions in the original arrangement: $A \downarrow BCD \downarrow EFG \downarrow HI \downarrow J$. Arrangement (c) was produced by two overlapping inversions: $A \downarrow BCDEF \downarrow GHIJ \rightarrow AFE \downarrow DCBGHI \downarrow J \rightarrow$ (c).

22-11. The sequence in the darkly colored chromosome can be explained as resulting from three inversions, one nonoverlapping and two overlapping. The two break points in the nonoverlapping inversion occurred between $2A$ - $2B$ and between $7C$ - $7D$. The two breakpoints in the first of the overlapping inversions occurred between $8C$ - $8D$ and between $12C$- $12D$. The two break points in the second of the overlapping inversions occurred between $8D$ - $8E$ and between $16C$ - $16D$.

22-12. (a) If the 21/21 isochromosome arose through translocation during meiosis in one of the parental gonads and was transmitted to a zygote, it would also be necessary to postulate that the zygote develops phenotypically normal because the other parent contributed a nondisjunctional gamete lacking chromosome 21. That is, the explanation would rely on an extremely rare confluence of *two* different meiotic accidents (translocation, nondisjunction), each by itself individually rare. This seems highly improbable. On the other hand, only *one* accident is necessary to explain the origin of this translocation if we assume that it occurred during a mitotic cell division in the zygote at the "cleavage" stage soon after the zygote was formed. That is, the two 21 chromosomes of an early embryonic cell became attached to the same centromere, and this isochromosome was then passed on to all succeeding cells.

(b) All surviving offspring would be trisomic for chromosome 21 and therefore show Down syndrome.

22-13. (a) A translocation between chromosome 21 and one of the other chromosomes could be responsible (see text pp. 444-446). Thus, one parent in each family (i.e., II-1 and II-3) as well as one of the grandparents are probably translocation heterozygotes.

(b) The presence of a translocated 21 chromosome (e.g., 14^{21}) reduces the total number of chromosomes in cells of the translocation heterozygote to 45, since such individuals possess, for example, one normal 14 chromosome, one normal 21 chromosome, and the translocated 14^{21} chromosome, in addition to the other 42 chromosomes (21 pairs). (See text Fig. 22-22.)

(Reference: M.W. Shaw, 1962, *Cytogenetics*, 1:141-179.)

22-14. (a) One possibility is that this configuration is caused by a three-strand double crossover in a heterozygote for a paracentric inversion in which one crossover occurs within the inversion and the other occurs between the inversion and the centromere. In the diagram below, pairing between homologues and subsequent crossing over between indicated numbered strands, will lead to formation of a "loop" and an acentric fragment when the homologues separate at anaphase of meiosis I. (At anaphase of meiosis II, the centromeres of the "loop" chromosome will separate, producing a dicentric chromosome.)

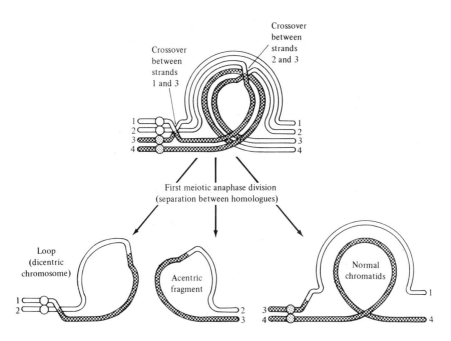

(b) As shown clearly in text Fig. 22-11, this can be explained by a four-strand double crossover in a heterozygote for a paracentric inversion, both crossovers occurring within the inversion.

22-15. Single crossovers within a paracentric inversion produce cells at anaphase I that show one dicentric and one acentric chromosome. Double crossovers are of three kinds: two-strand, three-strand, and four-strand in the ratio of 1:2:1 if chromatid interference is absent (see text p. 327). As shown in text Fig. 22-11, the consequences of each of these double crossovers in paracentric inversions will be different. If we summarize the frequencies of each kind of meiotic product we would expect the following:

		Meiotic Product		
Frequency	Crossover Type	Normal	One Dicentric, One Acentric	Two Dicentrics, Two Acentrics
.50	single		.500	
.125	2-strand double	.125		
.50 ⎰ .250	3-strand double		.250	
.125	4-strand double			.125
		.125	.750	.125

Thus, one eighth of the meiotic cells would appear normal at anaphase I, and the remaining seven eights would be divided into two abnormal types in a ratio of six cells carrying both a dicentric and acentric chromosome to one cell carrying two dicentrics and two acentrics.

22-16. (a) The egg nuclei should contain either a noncrossover rod-X or a noncrossover ring-X. The crossover products produce a dicentric chromosome that is eliminated in the polar bodies.

(b) The egg nuclei should contain either crossover or noncrossover ring-X or rod-X chromosomes.

(c) Nullo-X eggs should appear since a double bridge is formed to which all the crossover products are attached. Occasionally, as shown by Novitski, a centromere carrying an X-chromosome fragment (formed by breakage of the bridge) is drawn into the egg nucleus, thereby producing an aneuploid.

22-17. One possible approach to this problem is to note that if the *y* (*yellow*) and *sn* (*singed*) markers are on the ring-X chromosome, a female that is carrying this chromosome and a wild-type rod-X will produce mosaic mutant spots only when somatic crossing over occurs but not on elimination of the ring-X chromosome. On the other hand, when the *y* and *sn* markers are carried on the rod-X chromosome but not on the ring-X, mosaic mutant spots are produced both on somatic crossing over and elimination of the ring-X. A comparison of these two sets of results therefore may give some notion of the relative importance of ring-X chromosome elimination if we consider somatic crossing over to be similar in both experiments. In actuality, somatic crossing over probably differs, and more sophisticated experiments have been done by Brown, Walen, and Brosseau. (Somatic crossing over and elimination of ring-X chromosomes of *Drosophila melanogaster.* 1962, *Genetics,* **47**:1573-1579.)

22-18. (a) Translocation   (b) Translocation

22-19. To detect a translocation, observations can be made cytologically (e.g., see text Fig. 22-17) or a search can be made for specific genetic changes (the establishment of new linkages between previous independently assorting genes and the modification of old linkages into new independent assortments). Lethals would not have these consequences.

22-20. Each of the six males is heterozygous for a translocation involving two chromosomes and associated genes as follows:

(a) 2-3, carrying either Bl-D or Bl^+-D^+
(b) 3-4, carrying either D-ey or D^+-ey^+
(c) 2-4, carrying either Bl-ey or Bl^+-ey^+
(d) 2-3, carrying either Bl-D^+ or Bl^+-D
(e) 3-4, carrying either D-ey^+ or D^+-ey
(f) 2-4, carrying either Bl-ey^+ or Bl^+-ey

(Reference: Th. Dobzhansky, 1936. *Biological Effects of Radiation*, B.M. Duggar (ed.). McGraw-Hill, New York, pp. 1167–1208.)

22-21. A translocation between the X chromosome and the second chromosome bearing the *Bl* gene. (Note that the male offspring of matings with attached-X females inherit their X chromosome from their father.) Since *Bl* is now linked to the paternal X chromosome, all males will be *Bl*. The third chromosome gene, *D* and its wild-type allele, will assort independently of the other chromosomes.

22-22. A translocation between the Y chromosome and the third chromosome bearing the *D* gene. Since the female offspring of matings with attached-X females ($\widehat{XX}$Y) inherit the Y chromosome from their father (and receive the $\widehat{XX}$ from their mother), all attached-X females in these crosses will carry the paternal Y chromosome with its dominant *D* gene.

22-23. The F_2 females inherit their attached-X chromosomes from their mothers and therefore all have the yellow-body phenotype, whereas F_2 males inherit the wild-type yellow-body allele on their father's X chromosome. Appearance of autosomal traits would be expected as follows:

(a) In the F_1 males, the bw^+-e^+ alleles would be transmitted as a unit, and the bw-e alleles would also be transmitted as a unit, with *eyeless* and the sex chromosomes assorting independently of each other and of the brown-ebony unit. So for both F_2 female and male progeny, expected phenotypes would be in the ratio 1/4 bw^+-e^+; ey^+ (wild type) :1/4 bw^+-e^+; ey (eyeless):1/4 bw-e; ey^+ (brown-ebony):1/4 bw-e; ey (brown-ebony, eyeless).

(b) In the F_1 males, the e^+-ey^+ alleles would be transmitted as a unit, and e-ey alleles would also be transmitted as a unit, with *brown* and the sex chromosomes assorting independently of each other and of the ebony-eyeless unit. So for both F_2 female and male progeny, expected phenotypes would be in the ratio 1/4 e^+-ey^+; bw^+ (wild type):1/4 e^+-ey^+; bw (brown):1/4 e-ey; bw^+ (ebony eyeless):1/4 e-ey; bw (brown-ebony, eyeless).

(c) In the F_1 males, the bw^+-e^+-ey^+ alleles would be transmitted as a unit, and the bw-e-ey alleles would also be transmitted as a unit, assorting independently of the sex chromosomes. So for both F_2 males and females, phenotypes would be in the ratio 1/2 bw^+-e^+-ey^+ (wild type):1/2 bw-e-ey (brown-ebony, eyeless).

(d) The X chromosome and the e^+ allele would be transmitted as a unit (X-e^+), with *brown* and *eyeless* genes assorting independently of each other and of the X-e^+ unit in F_2 males. In the ($\widehat{XX}$Y) F_2 females, the Y chromosome and the e allele would be inherited as a unit (Y-e), with *brown* and *eyeless* assorting independently as they do in the males. So the F_2 males would be phenotypically in the ratio 1/4 wild type:1/4 brown:1/4 eyeless:1/4 brown eyeless, and the females would be in the ratio 1/4 ebony:1/4 brown-ebony:1/4 ebony eyeless:1/4 brown-ebony eyeless.

(e) In the F_1 males, the X chromosome and the bw^+ and ey^+ alleles would be transmitted as a unit (X-bw^+-ey^+), and the Y chromosome and the bw and ey alleles would also be transmitted as a unit (Y-bw-ey), with *ebony* assorting independently. Thus, phenotypes of the F_2 male progeny would be in the ratio 1/2 wild type:1/2 ebony, and F_2 females would be in the ratio 1/2 brown eyeless:1/2 brown-ebony eyeless.

(f) In the F_1 males, the wild-type X chromosome and the three wild-type autosomes would be transmitted as a unit. The F_1 male Y chromosome, together with the three recessive-bearing autosomes would also be transmitted as a unit. Thus, all F_2 males would be phenotypically wild type, and all F_2 females would be brown-ebony, eyeless.

(Reference: J.T. Patterson, W. Stone, S. Bedichek, and M. Suche, 1934, *Am. Natur.*, **68**:359–369.)

22-24. If we symbolize the mutant allele *brachytic* with *b* and the normal allele of *brachytic* with *B*, the translocated chromosomes with *T*, and the nontranslocated chromosomes with *t*, then the F_1 backcross can be represented as B/b $T/t \times b/b\ t/t$.

If *brachytic* segregated independently of the translocation, we would expect equal proportions for the following four groups of backcross progeny:

B/b T/t	B/b t/t	b/b T/t	b/b t/t
wild-type	wild-type	brachytic	brachytic
translocation	nontranslocation	translocation	nontranslocation
heterozygote	homozygote	heterozygote	homozygote
(semisterile)	(fertile)	(semisterile)	(fertile)

Since the data clearly indicate no equality in proportions between these phenotypes, it seems likely that the *B* gene is linked to a chromosome involved in the translocation, that is, a reciprocal translocation occurred between the chromosome carrying *B* and a nonhomologous chromosome:

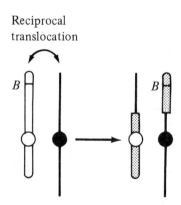

Reciprocal
translocation

Thus, the semisterile progeny (translocation heterozygotes) produced by crossing a plant carrying the above translocation to a homozygous *brachytic* stock can be diagrammed as follows:

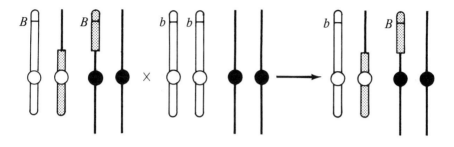

The backcross of these *B/b* translocation heterozygotes is therefore of the type:

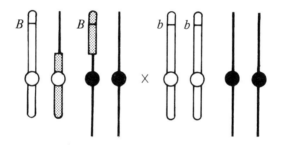

Thus, most of the "balanced" gametes produced by the translocation heterozygote are as follows:

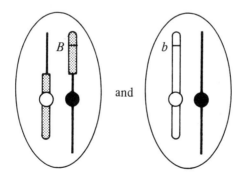

and

Upon fertilization, these gametes produce wild-type semisterile translocation heterozygotes and brachytic fertile nontranslocation homozygotes. The exceptional backcross progeny arise from crossing over in a paired section of the translocation heterozygote, which can be diagrammed as follows (noncrossover chromatids omitted for simplicity):

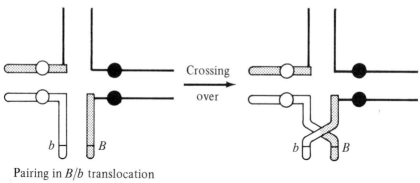

Crossing over

Pairing in *B/b* translocation
heterozygote

Upon *alternate* segregation, such crossover events would then give rise to gametes of the following constitution:

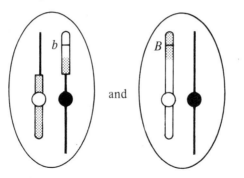

and

It is apparently these gametes, upon fertilization by the *brachytic* stock bearing nontranslocated chromosomes, that produce the infrequent brachytic, semisterile translocation heterozygotes and the wild-type, fertile, nontrans-location homozygotes.

Based on this explanation, the answers are therefore as follows:

(a) Equal proportions for each of the four phenotypes.

(b)   parentals = 334 + 279 = 613
      recombinants — 27 + 42 = 69
                  Total 682

Brachytic - translocation linkage distance = 69/682 = 10.1 percent.

(Reference: R.A. Brink and D.C. Cooper, 1931, *Genetics*, **16**:595–628.)

22-25. The parents are of two kinds: (1) phenotypically wild type and carry-ing a translocation (+ + T); (2) phenotypically liguleless-virescent and not carrying a translocation (*lg v* +). The semisterile F$_1$ is therefore heterozygous

$+ + T / lg v +$ and, if assortment were independent between these three factors, would be expected to produce eight types of backcross progeny in equal proportions. The fact that the observed proportions are unequal indicates that the three factors are linked (we already know that lg and v are on the same chromosome), and linkage distances can be calculated for each pair of factors as the frequency of recombinants:

$$lg - v = (126 + 124 + 14 + 18)/588 = 47.95 \text{ percent}$$
$$v - T = (25 + 13 + 14 + 18)/588 = 11.90 \text{ percent}$$
$$lg - T = (126 + 124 + 25 + 13)/588 = 48.97 \text{ percent}$$

The linkage order therefore seems to be $lg - v - T$, and the $lg - T$ distance therefore should be increased by twice the frequency of double crossovers $(14 + 18)$, or $lg - T = 59.85$ percent.

22-26. One possible approach is to call one plant X and the other plant Y. Each plant may then be considered to have two chromosome complexes with arms numbered as follows (each complex going to an individual pole during alternate segregation):

Plant X, complex X_1: $1 \cdot 2, 3 \cdot 4$
complex X_2: $1 \cdot 3, 2 \cdot 4$ ⎫ ring of 4

Plant Y, complex Y_1: $1 \cdot 2, 3 \cdot 4$
complex Y_2: $1 \cdot 4, 2 \cdot 3$ ⎫ ring of 4

A cross between plants X (X_1/X_2) and Y (Y_1/Y_2) therefore will produce the following offspring: X_1Y_1 (2 bivalents); X_1Y_2 (ring of 4); X_2/Y_1 (ring of 4); X_2/Y_2 (ring of 4).

22-27. As in the above answer, we can call one plant X and the other Y, and use the $1 \cdot 2, 3 \cdot 4, 5 \cdot 6$ system for naming the chromosome arms in each complex. For example, if we designate the chromosome arms of complex X_1 as $1 \cdot 2, 3 \cdot 4, 5 \cdot 6$, then, since X times Y produces 1/4 plants showing three bivalents during meiosis, we can assume that one of the Y complexes, for example, Y_1, is also $1 \cdot 2, 3 \cdot 4$, $5 \cdot 6$. However, in order to obtain an X X Y plant with a ring of four and one bivalent, the other chromosome complex in Y (Y_2) can be designated so that it differs from X_1 in two chromosomes (thereby producing the ring of four) but is identical to X_1 in one chromosome (thereby producing the bivalent). For example, if Y_2 is $1 \cdot 2, 3 \cdot 5, 4 \cdot 6$, then X_1/Y_2 will produce the meiotic figure shown at the top of the next page (for diagrammatic simplicity, individual chromatids are omitted).

To obtain the given proportion of rings of six chromosomes from the X X Y cross, we need an X_2 complex that differs in all three chromosomes from both Y_1 $(1 \cdot 2, 3 \cdot 4, 5 \cdot 6)$ and Y_2 $(1 \cdot 2, 3 \cdot 5, 4 \cdot 6)$. It is clear that if X_2 is $2 \cdot 3, 4 \cdot 5, 6 \cdot 1$,

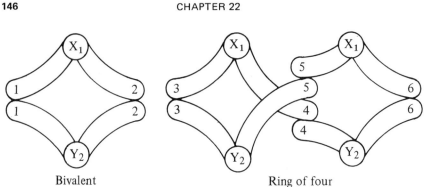

Bivalent   Ring of four

for example, it will yield progeny showing rings of six when combined with either Y_1 or Y_2 (and also with X_1). Thus, a cross between the two plants $(X_1/X_2 \times Y_1/Y_2)$ produces one fourth with three bivalents (X_1/Y_1); one fourth with a ring of four and one bivalent (X_1/Y_2); and one half with a ring of six $(1/4\ X_2/Y_1 + 1/4\ X_2/Y_2)$.

22-28.

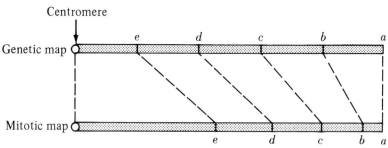

23

Gene Mutation

23-1. Breed the mouse with a homozygous stock of normal mice and observe the offspring. An effect caused by a dominant gene will appear in the F_1; a recessive effect will not. If the effect does not appear (i.e., if it is recessive or if it had been previously environmentally induced), inbreed the F_1 and observe the F_2. The F_2 ratios will indicate one- or two-gene segregation, that is, gene interaction and, if large enough numbers are raised, even three-gene segregation. Environmental induction of the trait is not expected to be correlated with genotypic ratios.

23-2. One important clue distinguishing between the two possibilities would be the frequency with which the new phenotype appears. Recombination frequencies are much higher than mutation frequencies, and traits caused by the former would be expected to occur with significantly greater frequency than those caused by the latter. Also, if a recombinational event is suspected, one could look for the appearance of novel chromosomal translocation complexes (see, e.g., text p. 450).

23-3. If the effect is caused by a deletion, then half the male offspring of this deletion heterozygote might be lethal since they are missing part of their X chromosome, that is, the offspring of this female would be in the ratio of 2 ♀:1 ♂. If the effect is caused by a point mutation, no such change in normal sex ratios is expected. Another genetic test is to observe whether the degree of recombination is reduced between genes in the *miniature* genotype. A cytological test is to observe the salivary chromosomes of the larvae produced by the *miniature* female. Half the female larvae produced will be heterozygous for this effect, and if it is a deletion, it may show up as a "buckling" caused by missing salivary bands in the *miniature* region.

23-4. If the disease in this pedigree is caused by transmission of a sex-linked recessive gene through previous generations, then female I-2 must be a heterozygous carrier. Note, however, that all six of her sons are free of the disease, a fairly low probability of $(1/2)^6$. It therefore seems more likely that a recent mutation is responsible for the appearance of the disease, and it occurred in one of the gametes that produced female II-3.

(Reference: K. Itiaba, M. Banfalvi, J.C. Crawhall, and J.G. Mongeau, 1973, *Am. J. Hum. Genet.*, **25**:134-140.)

23-5. In this cross, male offspring inherit the paternal X chromosome. If some of the paternal X chromosomes are lethal, males carrying these chromosomes will not appear. However, there would be no way of identifying the absence of such males by their lethality since their frequency would not be expected to be high. (If the frequency of sex-linked recessive lethals in the paternal sperm was very great, this of course would cause a large distortion in the sex ratio produced in this mating and could probably be detected.)

23-6. Plate an *arg-2* strain on a medium that lacks arginine but is supplemented with pyrimidine. Mutants that are both *pyr-3* and *arg-2* will now be able to grow on this medium. (Revertants of *arg-2⁻* to *arg-2⁺* will also grow on this medium but can be separated from the *pyr-3 arg-2* mutants by the fact that they can grow easily on a medium lacking pyrimidine.)

(Reference: J.L. Reissig, 1960, *Genetic Res.*, **139**:356-374.)

23-7. Isolation of the double mutant, *arg⁻leu⁻*, can be accomplished in a two-step selective process with penicillin. The first step is to subject a large number of bacteria to a complete supplemented medium in which penicillin is present but arginine is absent. Only the *arg⁻* mutants will survive since they are the only bacteria that are not dividing and are therefore not subject to the killing action of penicillin. These mutants can then be isolated and plated on arginine-supplemented medium without penicillin. The second step is to plate large numbers of the above *arg⁻* strain on a complete supplemented medium treated with penicillin in which leucine is absent. The only surviving bacteria will then be *arg⁻* bacteria that are also *leu⁻*. These double mutants will grow on medium supplemented with arginine and leucine.

23-8. Grow the streptomycin-resistant cells on streptomycin-free medium. Those colonies that grow on this medium are presumably streptomycin-independent and can be retested for streptomycin resistance on streptomycin medium.

23-9. (a) All progeny should be arginine-independent.

(b) In independent assortment, the *arginine* mutation will segregate into the same spore as the wild-type allele of the suppressor gene about one fourth of the time, that is, this proportion of the spores will produce colonies that are arginine-dependent.

(c) The proportion of spores carrying *arginine* and the wild-type allele of the suppressor gene will be reduced to one half the recombination frequency, that is, the expected frequency of arginine-dependent progeny is 10 percent.

23-10. (a) Since chondrodystrophic dwarfism is usually caused by a dominant gene (see text pp. 108-109), the cause for the sudden appearance of this trait is probably mutation.

(b) The fact that a parent is a heterozygote does not mean that 50 percent of his or her offspring must inherit the gene. Each child has a 1/2 chance of inheriting it, and the possibility that three children who do not carry the gene will be produced is $(1/2)^3$; that is, one out of eight three-child families with one chondrodystrophic parent will show, by chance, no dwarfism among the offspring.

(c) Since albinism is mostly a recessive trait, it has probably arisen here by a mating between heterozygotes with normal phenotypic appearance.

23-11. Calculation of mutation rates for dominant traits is explained on text p. 465. In this case $2/322{,}182 \times 1/2 = .31$ per 100,000 gametes.

(Reference: C.E. Blank, 1960, *Ann. Hum. Genet.*, **24**:151-164.)

23-12. $U = Nu = 50{,}000\ (.00004) = 2$ mutations per gamete

23-13. The fact that all white facets are of split phenotype in variegated sections of the eye, indicates that the heterochromatic position effect on *split* ($spl^+ \rightarrow spl^-$) can "spread" to *white* ($w^+ \rightarrow w^-$), but when it does (as evidenced by the appearance of white-eye facets), the position-effect modification of *split* must have already occurred. The reason for this, of course, is that the *split* gene is closer to the heterochromatin than *white*, and the spread of the position effect begins at the heterochromatin. Thus, $split^+$ can be modified by position effect independently of *white* and produce split red-eyed facets as well as split white-eyed facets. However, when the two genes are equidistant from the heterochromatin, variegation at the *white* locus can now occur independently of *split* (and vice versa), causing the appearance of some white facets that are not of split phenotype.

(Reference: W.K. Baker, 1963, *Am. Zool.*, **3**:57-69.)

23-14. (a) The mean and variance for the samples from independent cultures are 105.9 and 3067.25, respectively. For the samples from the single culture these values are 131.2 and 161.89, respectively.

(b) It is obvious that the high variances between the independent cultures indicate much greater real differences between them than do the small chance variances between the samples taken from a single culture.

(c) As indicated on text pp. 476-477, such results indicate that the mutations involved are chance events (preadaptive) and do not arise as a result of exposure to the selective medium (postadaptive).

(Reference: M. Demerec, 1948, *J. Bacteriol.*, **56**:63-74.)

23-15. Temperate phage resistance occurs through a change of the infecting phage to the symbiotic prophage form (lysogeny), which then inhibits further infection of the same phage type (see text pp. 359ff.) In other words, resistance of bacteria toward a temperate phage would arise by *contact* with the phage.

The Luria-Delbruck experiment would therefore have mistakenly indicated that mutation is *postadaptive*, unless it was known that the phage resistance was caused by lysogeny. (The same mistaken conclusions could have been drawn if phage λ had been used by the Lederbergs.)

23-16. This experiment can be done by growing a "lawn" of F^+ auxotrophic *E. coli* cells on a petri dish used as a "master plate." Replicas taken from this master plate by means of oriented velvet blocks (see text Fig. 23-15) can then be used to touch petri dishes containing lawns of F^- auxotrophic cells that are on *unsupplemented* medium and therefore unable to grow. By proper selection of F^+ and F^- strains, only *recombinant* prototrophs (formed by Hfr cells) could grow on this medium. For example, the F^+ strain would be $A^- B^- C^+ D^+$ and the F^- strain would be $A^+ B^+ C^- D^-$, and the medium for the replica plates would be $A^- B^- C^- D^-$. Thus without the transfer of chromosomal material from the F^+ to the F^-, *no* growth could occur on the replica plates. Only when F^+ cells have become Hfr and have transferred $C^+ D^+$ genes to the F^- cells could growth occur. Therefore, the appearance of growing colonies on the replica plates indicates the origin of Hfr strains that can then be traced to their origin on the master plate. (An actual experiment of this type was reported by Jacob and Wollman in 1956, *C. R. Acad. Sci. Paris*, **242**:303.)

23-17. Demerec's findings indicated that the stepwise improvement of bacteria resistant to penicillin was caused by a step-wise selection process of *different* penicillin-resistant mutations. Since each mutation occurs in a frequency of about 10^{-7}, the chances of finding a bacterium in which two different mutations had arisen simultaneously would be exceedingly small, $10^{-7} \times 10^{-7} = 10^{-14}$. Similarly, a triple-mutant bacterium could be found only in a frequency that would involve raising about 10^{21} cells ($10^{-7} \times 10^{-7} \times 10^{-7} = 10^{-21}$). It is therefore obvious why the isolation of bacteria resistant to high concentrations of pencillin must be a step-wise process; single mutants must first be selected, these single mutants must be raised and multiplied, and then a second-step mutation must be selected among these single mutants, and so on. Thus, the chances for a penicillin-sensitive bacterial population to develop immediate resistance to a high concentration of penicillin is very small, since many step-wise mutations are involved. Treatment of bacterial infection using penicillin should therefore begin with as high a dose of penicillin as possible.

(Reference: M. Demerec, 1948, *J. Bacteriol.*, **56**:63-74. See also M. Demerec, 1945, *Proc. Nat. Acad. Sci.*, **31**:16-24.)

23-18. A strain carrying a *single* auxotrophic mutation will grow on minimal medium only when the medium is supplemented with the specific substance affected by the mutation but will not grow on minimal media to which substances other than the necessary one has been added although the strain is prototrophic for these other substances. Thus strain 1, for example, is obviously auxotrophic for substance C, and prototrophic for substances A, B, and D. Strain 4, on the other hand, can grow on all minimal media, irrespective of which particular substance is added (or missing), and must therefore be prototrophic for *all* substances or wild type. Strain 2, however, does not grow at all on any of these supplemented media, indicating that it carries a mutation that cannot be corrected by any of the four supplementary substances; that is, it is auxotrophic for a substance other than A, B, C, and D (but prototrophic for A, B, C, and D, since it carries only a single auxotrophic mutation). On this basis, prototrophic (+) and auxotrophic (−) designations for the six strains are as follows:

		Substance		
	A	B	C	D
1	+	+	−	+
2	+	+	+	+
3	+	−	+	+
Strains 4	+	+	+	+
5	−	+	+	+
6	+	+	+	−

24

Induced Genetic Changes and DNA Repair Mechanisms

24-1. Since 12 percent mutation is produced by 4000 r of x-rays, a strict linear relationship between mutation rate and dosage yields 3 percent mutation (12/4) for each 1000 r. The answers are therefore as follows:

      (a) 3 percent   (b) 6 percent   (c) 15 percent   (d) 18 percent

24-2. (a)

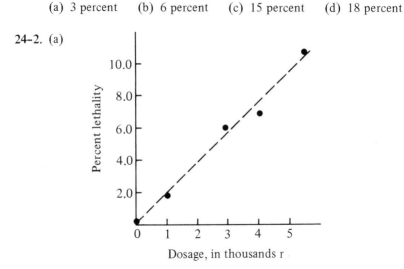

     (b) Considering x-ray dosage as X, and percent lethality as Y, then $\Sigma X = 13{,}000$; $\Sigma Y = 25.66$; $\Sigma X^2 = 51{,}000{,}000$; $\Sigma XY = 100{,}730$. The regression coefficient (text p. 262) is then calculated as:

$$b = \frac{\Sigma XY - (\Sigma X \, \Sigma Y / N)}{\Sigma X^2 - (\Sigma X)^2/N} = \frac{100{,}730 - 333{,}580/5}{51{,}000{,}000 - (13{,}000)^2/5}$$

$$= \frac{100{,}730 - 66{,}716}{51{,}000{,}000 - 33{,}800{,}000} = \frac{34{,}014}{17{,}200{,}000} = .002$$

    (c) $Y = \overline{Y} + b(X - \overline{X}) = 5.1320 + .002\,(6000 - 2600) = 11.93$ percent.

(Reference: M. Demerec, 1938, *Radiology*, **30**:212-220.)

24-3. (a)

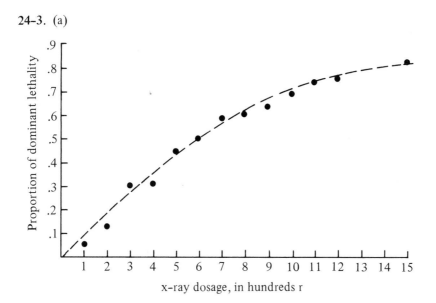

x-ray dosage, in hundreds r

(b) The proportion of dominant lethality is 1 — the relative survival of eggs. The correlation is therefore based on the values in columns **a** (X) and **c** (Y) in the following table:

a	b	c	a	b	c
X-ray Dosage, r	Relative Survival of Eggs	Proportion of Dominant Lethality	X-ray Dosage, r	Relative Survival of Eggs	Proportion of Dominant Lethality
0	1.000	0	700	.424	.576
100	.940	.060	800	.399	.601
200	.829	.171	900	.366	.634
300	.700	.300	1000	.319	.681
400	.687	.313	1100	.277	.723
500	.544	.456	1200	.255	.745
600	.500	.500	1500	.195	.805

$\Sigma X = 9300$; $\Sigma Y = 6.565$; $\Sigma XY = 5815.8$; $\Sigma X^2 = 8{,}750{,}000$; $\Sigma Y^2 = 3.963$

The correlation coefficient (text p. 260) is then calculated as follows:

$$r = \frac{\Sigma XY - (\Sigma X \, \Sigma Y / N)}{\sqrt{(\Sigma X^2 - [(\Sigma X)^2/N])(\Sigma Y^2 - [(\Sigma Y)^2/N])}}$$

The value of r we obtain is .96, indicating a very high positive correlation between x-ray dosage and dominant lethality.

24-4. Any process that produces homozygosity in a diploid organism can be used to detect the presence of recessive lethals. In *Paramecium* this is autogamy (see text p. 32), and a detailed description of the methodology used is given by R. F. Kimball and S. W. Perdue in 1962, *Genetics*, 47:1595-1607.

24-5. No, since the temperature effect depends on increasing the amount of oxygen in the tissues. In the absence of oxygen no such increase is expected.

24-6. Assuming that replication of an irradiated DNA strand can occur under some circumstances, the complementary strand would probably show a decrease in the frequency of A-A nearest neighbors since thymine dimerization can be expected to upset normal hydrogen bonding and base-pairing relationships. Also, since the repair of UV damage in the cell may involve removal of thymine dimers, a loss of A-A nearest neighbors complementary to such strands might be expected.

24-7. As in x-ray mutagenesis, translocations are generally caused by two mutational events (two chromosome breaks) occurring simultaneously and are therefore exponential in frequency compared to the occurrence of only one-mutation events (lethals) that follow linear relationships.

24-8. (a) Depending on which nucleotides are affected and on whether the mutational errors occur during replication or incorporation (see text Fig. 24-14), a variety of changes are possible. Thus, if the normal keto form of BU is incorporated in place of thymine, a change to its rare enol form at any subsequent DNA replication will enable it to act as cytosine and pair with guanine. At the next DNA replication that guanine nucleotide will pair with a cytosine nucleotide, and an A-T pair will thus have been changed to a G-C pair (or a T-A pair will have been changed to a C-G pair). On the other hand, if the rare enol form of BU is incorporated in place of cytosine, it will most likely revert to its normal keto form at subsequent DNA replications and pair as thymine does, so that a G-C pair will have been changed to A-T (or C-G to T-A).

(b) 2-aminopurine (AP) is a base analogue of adenine (see text Fig. 5-10) that normally pairs with thymine. In its rare tautomeric form, AP behaves like guanine and pairs with cytosine. If AP is incorporated as adenine, subsequent tautomeric changes can cause transitional mutations of A-T to G-C, or T-A to C-G. On the other hand, if AP is incorporated in its rare form as guanine, reverse changes will occur: G-C to A-T or C-G to T-A (see text p. 493).

(c) Nitrous acid changes adenine to hypoxanthine, which then behaves like guanine and pairs with cytosine. This will cause transitions of A-T to G-C and T-A to C-G. Nitrous acid also changes cytosine to uracil, which then behaves like thymine and pairs with adenine. This will cause transitions of C-G to T-A and G-C to A-T (see text Figs. 24-15 and 24-16).

(d) Hydroxylamine modifies cytosine, so that it can pair with adenine rather than with guanine (see text Fig. 24-17). This will cause transitions of G-C to A-T and C-G to T-A.

(e) Ethyl methanesulfonate alkylates guanine, so that it can pair with thymine rather than with cytosine (see text Fig. 24-18). This will cause transitions of G-C to A-T and C-G to T-A.

24-9. According to text Fig. 24-14c, the A → G transition occurs when BU has been incorporated into DNA in its common keto form as an analogue of thymine. Subsequent changes of BU to its rare enol form (acting as an analogue of cytosine) will cause it to pair with G, and therefore lead to the A → G transition. Since a number of DNA replications may take place before BU assumes its enol form, the presence of BU in the medium is not necessary for this transition to occur. On the other hand, the G → A transition only occurs when BU in its rare enol form is incorporated to pair with guanine, and then subsequently reverts to its keto form and pairs with adenine (text Fig. 24-14d). The G-C → A-T transition is therefore not able to proceed without the presence of BU in the medium.

24-10. ϕX174 has single-stranded DNA, and therefore the occurrence of a mutation on this strand will produce a pure clone. Phage T4 has double-stranded DNA, and the occurrence of a mutation on one strand will leave the other one wild type. Thus the replication of a single T4 mutant can produce mixed colonies.

(Reference: I. Tessman, 1959, *Virology*, 9:375–385.)

24-11. Since hydroxylamine does not cause reversions of mutations induced by the incorporation of the enol form of 5-bromouracil or of mutations induced by itself, we can assume that these BU mutations are identical to the hydroxylamine mutations, that is, G-C to A-T. Since hydroxylamine can cause reversions of mutations induced by the incorporation of 2-aminopurine, we can sssume that the aminopurine mutations produce G-C pairs from A-T pairs, that is, A-T to G-C (or T-A to C-G).

24-12. Transversions. The fact that base analogues and hydroxylamine, which are transition-producing mutagens, would not cause reversion of the two *rII* mutations indicates that the mutations are probably not transitional (C ↔ T or A ↔ G). The lack of reversion by proflavin (or spontaneous reversion) indicates that the given *rII* mutations are not simple insertions or deletions. Thus, by the process of elimination, these two mutations probably represent transversions that are of the mutant A-T type, and are then changed (by thymine deprivation) back into C-G or T-A.

(Reference: M.D. Smith, R.R. Green, L.S. Ripley, and J.W. Drake, 1973, *Genetics*, 74:393–403.)

24-13. Removal of the mismatched guanine nucleotide at the primer terminus is a property of the 3′ → 5′ exonuclease activity of DNA polymerase (proofreading function) and would be expected to be defective in "mutator" genotypes and more effective than wild type in "antimutator" genotypes. Thus, the polymerase of genotype *A* can be considered a mutator since almost none of the mispaired guanine nucleotides is removed, and genotype *C* is an antimutator that removes these mismatches even more rapidly than wild type (genotype *B*).

25

Genetic Fine Structure

25-1. (a) The reversions to wild type appear too frequently to be explained by normal mutation rates. Also, the appearance of wild-type eyes in each case is always accompanied by the loss of *al*; this is obviously not merely a "double" mutation.

(b) Through recombination, as follows:

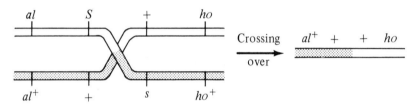

(c) The order is *al - S - s - ho*. Note that if the order were *al - s - S - ho*, a single crossover would produce wild-type eye phenotypes that were mutant for *al* rather than for *ho*:

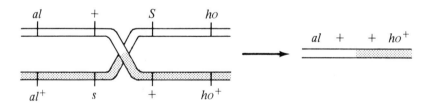

In order for wild-type eye phenotypes to appear that were mutant for *ho*, the crossover events for the *al - s - S - ho* gene order would have to be as follows:

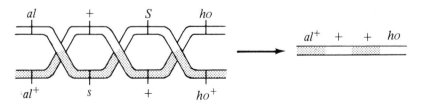

Since four recombinants of genotype $al^+ + + ho$ were observed among the offspring, it would be difficult to explain their origin as arising from relatively "common" triple crossovers in the gene order al - s - S - ho when single crossovers for this order $(al + + ho^+)$ are absent.

(Reference: E.B. Lewis, 1941, *Proc. Nat. Acad. Sci.*, **27**:31–35.)

25-2. (a) Single: $a^+ b^+$.   Doubles: $ab^+, a^+ b$.   Triple: ab.
    (b) Single: ab.   Doubles: $ab^+, a^+ b$.   Triple: $a^+ b^+$.

25-3. (a) Single: $a^+ b$.   Doubles: $ab, a^+ b^+$.   Triple: ab^+.
    (b) Single: ab^+.   Doubles: $ab, a^+ b^+$.   Triple: $a^+ b$.

25-4. In all crosses the single crossover classes can be recognized quite easily as being the most numerous. Note that in a cross *ylo pan a trp*$^+$ × *ylo*$^+$ *pan b trp*, the single crossover class that is wild type for *pan* is *ylo*$^+$ *trp*$^+$ if the order is *ylo* - *pan a* - *pan b* - *trp*, and is *ylo trp* if the order is *ylo* - *pan b* - *pan a* - *trp*. On this basis, the order of loci is *ylo* - *pan 25* - *pan 20* - *pan 18* - *trp*.

(Reference: M.E. Case and N.H. Giles, 1958, *Cold Sp. Harb. Symp.*, **23**:119–134.)

25-5. (a) The analytical procedure that can be followed is the same as that used for complementation mapping of the data in text Fig. 25–8 (pp. 510–512) and the deletion mapping used by Benzer (text p. 514). The two main rules for this kind of mapping are (1) mutations or deletions that *cannot* complement each other or recombine with each other to produce wild type *affect* the same product or the same region (they "overlap"); (2) mutations or deletions that *can* complement each other or recombine with each other to produce wild type *do not affect* the same product or same region (they do not "overlap"). In the present case, each of the *histidine* mutations can be considered to affect a region of the genetic map, with mutations that can recombine to form wild type affecting different regions and mutations that cannot recombine to produce wild type affecting the same region. Thus mutation A affects a different region than does B or D, and C affects a different region than does D, whereas B, C, and D all affect the same region ("overlap"). Two possible maps that can satisfy these conditions are therefore as follows:

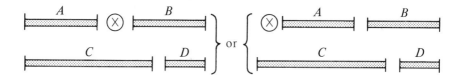

(b) Since the point mutation is in the same region as C, but in a different region than A, B, and D, its position in the above maps can be indicated by the symbol ⊗.

25-6.   Complementation Regions

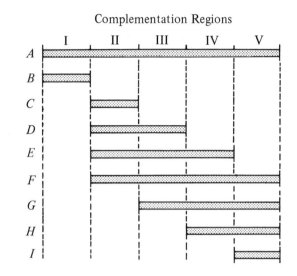

(Reference: D.G. Catcheside, 1965, *The Fungi*, Vol. 1, G.C. Ainsworth and A.S. Sussman, eds. Academic Press, New York, pp. 659-693.)

25-7.

	1	2	3	4	5	6
1	0	+	+	0	+	0
2	+	0	+	0	0	+
3	+	+	0	+	0	0
4	0	0	+	0	+	+
5	+	0	0	+	0	+
6	0	+	0	+	+	0

25-8. Two mutations in the *trans* position that are unable to complement each other to produce active phage can be presumed to be on the same cistron. Therefore the A cistron only contains rV and rW; the B cistron has rU, rX, rY and rZ.

26
Genetic Control of Proteins

26-1. Each protein molecule is composed of four closely related or identical polypeptide chains.

26-2. The dipeptides can be arranged in "overlapping" order:

                ala - val
                     val - his
                          his - ser

The order is therefore ala - val - his - ser.

26-3. Based on the analytical system described on text pp. 532–533 (see also Fig. 26-10), the most consistent way of arranging the amino acids of the given fragments is as follows:

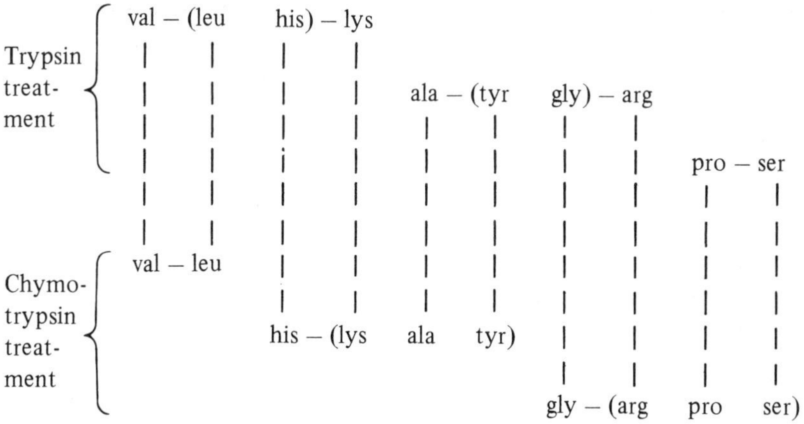

The order of these amino acids in the polypeptide is therefore

H_2N val - leu - his - lys - ala - tyr - gly - arg - pro - ser COOH

26-4. (a) According to the analytical method discussed on text pp. 527–528, it is obvious that arginine is at the end of the pathway (since all the different *arg* mutations can use it for growth) and glutamic semialdehyde is at the beginning of the pathway (since very few *arg* mutations can utilize it for growth). Intermediate to these substances are ornithine and citrulline, with the former earlier in the pathway than the latter, since ornithine can be used by fewer *arg* mutations than citrulline. In summary, the arginine biosynthetic pathway and the listed mutations that affect its synthesis can be pictured as follows (dashed lines indicate a mutant block):

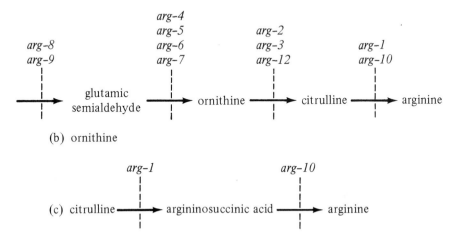

(b) ornithine

(c) citrulline

(Reference: A.M. Srb, and N.H. Horowitz, 1944. *J. Biol. Chem.*, **154**: 129-139.)

26-5. Mutant strain X^B is apparently unable to provide intermediary products that would allow growth of either strains X^A or X^C, and the block caused by strain X^B must therefore be early in the metabolic pathway. On the other hand, mutant strain X^A can provide intermediate substances that can be used for growth by both X^B and X^C, indicating that the metabolic block in strain X^A must be toward the end of the pathway. X^C can provide a substance necessary for the growth of X^B whose metabolic block is earliest in the pathway but cannot provide a substance that can be used by X^A whose block is apparently later in the pathway. Thus, the sequence in which the three mutant strains block the pathway to amino acid X is in the order: $X^B \rightarrow X^C \rightarrow X^A \rightarrow$ amino acid X.

27

Information Transfer and Protein Synthesis

27-1. There would be no relationship between the labeling of the particular peptides in the recovered hemoglobin molecules and the time after which the radioactive label was introduced. Instead, all peptides would be randomly labeled at time T_2 in Figure 27-10 in the text.

27-2. (a) No.   (b) No.   (c) Yes.

27-3. Their transfer and ribosomal RNAs might be alike, but their messenger RNAs would be expected to differ.

27-4. Penicillinase mRNA is long lived.

(Reference: H. Harris and L.D. Sabath, 1964, *Nature*, **202**:1078-1080.)

27-5. The correspondence between the peak frequency of radioactivity and the peak frequency of ribosomes immediately after the chase (text Fig. 27-14a) indicates that the ^{35}S-labeled amino acids are rapidly associated with ribosomes in the manufacture of polypeptides. Two minutes later, after the addition of ^{32}S, text Fig. 27-14b shows that many of the ^{35}S-labeled polypeptides have left the ribosomes, and the previous peak of radioactivity has disappeared from the ribosomal fraction. These findings clearly demonstrate that protein synthesis occurs on *E. coli* ribosomes, since radioactively labeled polypeptides associate with them in accord with expectations.

(Reference: K. McQuillen, R.B. Roberts, and R.J. Britten, 1959, *Proc. Nat. Acad. Sci.*, **45**:1437-1447.)

28
Nature of the Genetic Code

28-1. $2 \times 10^{12}/3 \times 500 \times 660 \cong 2{,}000{,}000$. It is, however, highly doubtful that an organism with a haploid complement of 2×10^{12} molecular weight of DNA such as man (see text Table 4-1, p. 47) makes 2 million different proteins. It seems more likely that some of the DNA is duplicated, producing the same RNA or protein in many DNA sections, such as has been observed for the production of ribosomal RNA in the nucleolar organizer regions (see text p. 637). It is also likely that some DNA, such as that in heterochromatin, has a function that is not involved in producing many different kinds of proteins, if any at all (see text pp. 85-86).

28-2. The relationship between the different kinds of nucleotides present in genetic material (N) and the numbers of different codon combinations that can be produced (C) can be formulated as $N^L = C$, where L stands for codon length; that is, the number of nucleotides in a codon sequence. Thus if DNA possesses only two kinds of nucleotides, codons that are five nucleotides long would give 32 possible combinations (2^5). Any codon of lesser length would be inadequate, for example, a four-nucleotide length codon provides only 2^4 or 16 possible combinations.

28-3. (a) Such a code was devised by Crick, Griffith, and Orgel (1957, *Proc. Nat. Acad. Sci.*, **43**:416–421), which had, of necessity, the additional quality that it was *nondegenerate*. That is, each of the 20 different amino acids had only one DNA triplet codon, and all of the remaining 44 triplets were "nonsense" and could not code for any amino acid. For example, if AAA = phenylalinine, CCC = leucine, GGG = valine, and a DNA strand had the sequence AAACCCGGG, then only the three given amino acids could be translated from this sequence no matter what the reading frame. The "intermediate" codons in this sequence, AAC, ACC, CCG, and CGG would be nonsense.

(b) A few of the theoretical arguments that oppose the existence of such a nondegenerate code are as follows:

1. Since only one codon specifies a particular amino acid in a nonde-generate code, *every* nucleotide change would lead to a significant change in a protein: either the substitution of a new amino acid,

or the appearance of a nonsense codon that would cause a deficient or inactive protein. Evolution of proteins by means of random mutation would therefore be fraught with danger and difficult to conceive.

2. Evolution of the genetic code itself must have occurred by trial-and-error, as did the evolution of other biological traits (see text Chapter 36), but a nondegenerate code could hardly have evolved since any codon changes would cause widespread serious effects. That is, the codons for all 20 amino acids would have to be specified at the immediate moment at which nucleotide-protein relationships were established if the code were to remain nondegenerate.

3. Presence of a nondegenerate code would indicate that protein synthesis occurs in the absence of a reading frame, and any nucleotide sequence bearing the proper codons can be translated into amino acids and proteins. There is, however, considerable evidence that not all nucleotide sequences are translated into proteins. In fact, nucleotide sequences of considerable length and complexity are not translated, and regulatory systems other than nondegeneracy must prevail. [Moreover, the evidence presented by Crick and co-workers (text pp. 571-573) shows the utilization of a reading frame.]

28-4. (a) The mRNA sequence begins AAU AGA · · ·. Using the genetic code dictionary in text Table 28-3, the amino acid sequence would therefore be asn - arg - ser - pro - leu - phe - cys.

(b) The deletion would cause the mRNA sequence to begin with the codon AAA, and the translated amino acid sequence would now be lys - glu - ala - leu - ser - phe.

28-5. The chances for one of the two different bases to be at any particular position is $1/2$, and the chances for the occurrence of any particular sequence of three nucleotides is $1/2 \times 1/2 \times 1/2 = 1/8$. Thus, if we restrict our choices to those codons containing U and/or A for the given amino acids (see text Table 28-3), the answers are as follows:

(a) UUU (phenylalanine) $= 1/8$
(b) AUU and AUA (isoleucine) $= 1/8 + 1/8 = 1/4$
(c) UUA (leucine) $= 1/8$
(d) UAU (tyrosine) $= 1/8$

28-6. On the basis of random nucleotide placement the expected frequency of each trinucleotide is obtained by multiplying the probability that each base will appear at that particular position in the codon, for example, AAA $= (.47)(.47)(.47) = 10.4$ percent; AAC $= (.47)(.47)(.53) = 11.7$ percent, etc. The expected amino acid percentages are therefore lys (AAA) $= 10.4$; asn (AAC) $= 11.7$; gln (CAA) $= 11.7$; his (CAC) $= 13.2$; thr (ACA, ACC) $= 24.9$; pro (CCA, CCC) $= 28.1$. This accords fairly well with the observed percentages of each amino acid.

(Reference: O.W. Jones, Jr., and M.W. Nirenberg, 1966, *Biochem. et Biophys. Acta*, **119**:400-406.)

28-7. (a) No. If the region were crucial for hormonal activity, we would expect very few amino acid changes to occur there and relatively more changes to occur in other regions.

(b) Yes, see Table 28-3 in the text.

28-8. (a) Glycine can be either GGU or GGC; if it is GGU then asp is GAU, ala is GCU and cys is UGU. If gly is GGC, then asp is GAC, ala is GCC, and cys is UGC on the basis of single nucleotide changes.

(b) Wild type in this case is the amino acid glycine (GG-). Therefore the answer is asp/cys (GA-/UG-) or ala/cys (GC-/UG-).

(c) The combination ala/asp (GC-/GA-) does not enable a GG- "wild-type" codon to be produced since neither of the recombining codons has G in the second position.

(Reference: J.R. Guest and C. Yanofsky, 1965, *J. Mol. Biol.*, **12**:793-804.)

28-9. Transitions, A ↔ G and T ↔ C, are represented by the change gly (GG-) ↔ asp (GA-), indicating that the DNA strand producing the mRNA must have undergone a change in the second position of the codon T ↔ C. Transversions, purines → pyrimidines, are represented by gly (GG-) → cys (UG-) and by asp (GA-) → ala (GC-), indicating changes in DNA of the type C ↔ A and T ↔ G, respectively.

28-10. No matter what codons are used, some amino acids are most likely represented in different positions by more than one type of codon. One possible set of codons involving a minimum of nucleotide substitutions is ala (GCA), glu (GAA, GAG), gln (CAA), pro (CCA), arg (CGA, CGU), leu (CUA), his (CAU), tyr (UAU), asp (GAU), gly (GGU, GGA), asn (AAU), lys (AAA), thr (ACU, ACA), ser (UCA), val (GUG), met (AUG).

(Reference: D. Beale and H. Lehmann, 1965, *Nature*, **207**:259-261.)

28-11. (a) Original: AAA - AGU - CCA - UCA - CUU - AAU - GCU. The changes probably involved a deletion of the first adenine base in the second codon and an insertion of G at the end of the next-to-the-last codon: AAA - GUC - CAU - CAC - UUA - AUG - GCU.

(b) Since translation proceeds from the 5′ end of mRNA to the 3′ end (see Chapter 27 and also text p. 576). the "reading frame" for this amino acid must begin with lysine at this end. If it began with alanine at this end, the amino acids shown in the mutant could not be obtained by a single nucleotide deletion and insertion.

(Reference: E. Terzaghi, Y. Okada, G. Streisinger, J. Emrich, M. Inouye, and A. Tsugita, 1966, *Proc. Nat. Acad. Sci.*, **56**:500-507.)

28-12. (a) The answer to this problem depends on noting the similarities between the initiator codon in polypeptide chain synthesis, AUG (see text Table 28-3 and p. 557), and the other codons at the N-terminal. In the normal polypeptide chain, the AUG codon comes before the threonine residue and codes for methionine (usually excised). In the four mutations, *A-D*, this initiating codon appears to have been modified by single nucleotide substitutions; that is, AUG → AUU, AUC, or AUA (isoleucine, mutation *A*), AUG → UUG or CUG (leucine, mutation *B*), AUG → AGG (arginine, mutation *C*), AUG → GUG (valine, mutation *D*). Each of the four mutations was therefore unable to produce a polypeptide chain since it no longer had an initiating codon at the N-terminal position. The reversion of each mutation occurred when an initiator codon was introduced before the formerly "mutated" initiator codon. (Evidence for this is seen in the fact that some of the revertants showed the presence of a methionine residue, e.g., met - ile - thr - glu - - -.)

(b) All of the four mutations were able to revert to the ala - gly - - - sequence by a single nucleotide substitution that changed the lysine codon (AAG) to the initiator codon (AUG). Thus, lysine is missing in these revertants, and the polypeptide chain begins with alanine.

(Reference: F. Sherman, J.W. Stewart, J.H. Parker, G.J. Putterman, B.B.L. Agrawal, and E. Margoliash, 1970, *Symp. Soc. Exp. Biol.*, 24:85-107.)

28-13. (a) Cysteine codons (UGU, UGC) can be coded by two tRNA "species," one bearing anticodon ACG (being wobbly, it fits either UGU or UGC) and the other ACA (fits only UGU). The tryptophan codon (UGG) pairs only with the tRNA species ACC, since C at the third codon position does not "wobble." The nonsense codon, UGA, like other nonsense codons does not have a tRNA species to pair with, and therefore no third-position wobble is involved. Thus, the wobble hypothesis enables a clear distinction between cysteine, tryptophan, and the UGA nonsense codons.

(b) No. To recognize only A at the third codon position, the tRNA would need to have U or I at its third anticodon position. However, either of these bases at this position could pair with at least one base other than A: U at the third anticodon position can also pair with G, and I can also pair with U and G. Note that Table 28-3 in the text has no amino acids with a codon restricted only to A at the third position.

(c) A suppressor of the *ochre* nonsense mutation (UAA) would be a tRNA with the anticodon AUU. According to the wobble hypothesis, the presence of U at the third position of this anticodon enables it to pair also with G and thereby suppress *amber* (UAG). On the other hand, a suppressor of *amber* such as tRNA with anticodon AUC can pair easily with UAG but does not have sufficient wobble at the third C position to pair with UAA (*ochre*).

28-14. The capacity for intragenic complementation might be expected to be related to the number of polypeptides that aggregate to form a protein molecule. The greater the number of aggregating polypeptides, the greater the chances that a mutational defect in one polypeptide chain can be compensated by a different but complementary mutation in another polypeptide chain which, together with the first, form a dimer or multimer. Since hemoglobin molecules consist of four polypeptide chains, 2α and 2β, whereas myoglobin consists of only one polypeptide, we would expect greater opportunities for complementation in the former than in the latter.

28-15. The nucleotide insertion (or deletion) causes a frameshift for the interval in which the nonsense mutation is found. Thus, the nonsense sequence UAG or UAA is not "read" as an individual codon within that interval but is read as part of two adjacent codons, for example,

$$\cdots \text{GCA UUG UAG ACC} \cdots \xrightarrow[\text{shift}]{\text{frame}} \cdots \text{GC AUU GUA GAC C} \cdots$$

(Reference: S. Brenner and A.O.W. Stretton, 1965, *J. Mol. Biol.*, 13:944-946.)

28-16. The "unknown" nonsense codon is UGA (*opal*), and the transition-producing mutagen, 2-aminopurine, causes the following single-step changes: UGA → CGA (arginine) or UGA → UGG (tryptophan). If the unknown nonsense codon is either UAA (*ochre*) or UAG (*amber*), then the reverse mutations are presumably UAA → CGA or UGG, and UAG → CGA or UGG. Note, however, that each of these latter reversions require two or three simultaneous changes, a much more improbable event than the single-step reversions from UGA. The recombinational evidence also shows that the nonsense codon is UGA, since it can recombine with *amber* to produce an amino acid codon:

(unknown)	U	G	A	
			✗	UGG (tryptophan)
(*amber*)	U	A	G	

Note that, if the unknown codon was either UAA or UAG, recombination with *amber* could not lead to the codon for an amino acid.

28-17. If the defect is caused by a frameshift mutation, it will not revert to a functional form when exposed to base-substitution mutagens such as hydroxylamine and 2-aminopurine, whereas nonsense mutations would revert. Also, nonsense mutations can be suppressed by *amber*, *ochre*, and *opal* suppressors whereas frameshift mutations are not suppressed in these strains.

28-18. Lengthening of the α hemoglobin polypeptide chain beyond its normal termination point indicates that the termination codon that ordinarily follows amino acid position #141 in Hb A has been changed in both genetic variants. In the case of Hb CS, the termination codon itself appears to have been modified by a nucleotide substitution, that is, UAA (termination) → CAA (gln), thereby permitting the nucleotide sequence beyond this point to be translated into amino acids. In the case of Hb W1, the nucleotide change responsible for changing the termination codon must have occurred at a prior nucleotide position since other amino acids in the polypeptide chain are also changed. Thus, if we assume that only one nucleotide was affected, the most reasonable hypothesis is that Hb W1 represents a frameshift mutation (nucleotide deletion) occurring at amino acid codon #139, which then leads to the misreading of subsequent codons including the termination codon UAA. Hemoglobin Wayne is now the most likely example of a frameshift mutation known in humans. One possible interpretation of these events in terms of messenger RNA nucleotide sequences is shown in the following figure (After J. Roth, 1974, Frameshift mutations. *Ann. Rev. Genet.*, 8:319–346.):

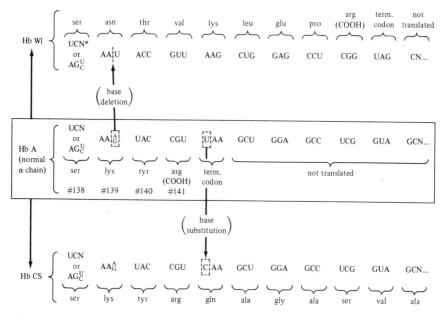

*N designates any of the four bases, A, U, G, or C.

29

Gene Regulation

29-1. If constitutive synthesis of β galactosidase in I^- mutants was caused by the presence of an internal inducer in this strain, then the introduction of an I^+Z^+ chromosome into an I^-Z^- recipient should have led to *continuous* constitutive enzyme synthesis (I^- inducer $+ Z^+$ gene). However, as shown by Pardee, Jacob, and Monod, constitutive synthesis in the I^- recipient lasted only for about 1 to 2 hours. After this period enzyme synthesis ceased, indicating the presence of an active repressor contributed by the I^+ donor.

29-2. This experiment indicates that β galactosidase synthesis begins anew after induction, since previous ^{35}S-labeled polypeptides are not found in its structure. Thus, the inducer does not cause the activation of enzyme polypeptides that are already present, but must act instead on a mechanism that enables new polypeptide synthesis.

29-3. (a) *Inducible synthesis.* Note that a normal repressor is produced (I^+) and the operon bearing A^+ has a normal operator (O^+). Therefore, acetylase production (A^+) will be affected by the repressor, and synthesis will be inducible.

(b) *Constitutive synthesis.* Note that the superrepressor (I^s) will prevent any A^+ synthesis on the operon carrying the normal operator (O^+). However, the I^s product cannot affect acetylase synthesis by the *lac* operon carried on the F' particle since that operon has an O^c mutation that will prevent repressor attachment, thereby leading to constitutive synthesis by its A^+ gene.

(c) *No synthesis.* Note that the only operon that has an A^+ gene is also carrying a normal operator gene (O^+), which is subject to the action of the superrepressor (I^s). This prevents acetylase production.

29-4. Constitutive enzyme synthesis would arise if repressor attachment to the operon is prevented, and transcription can proceed continuously. Repressor attachment, however, is to the operator site rather than the promoter site. The promoter site only affects RNA polymerase attachment, and promoter mutations can prevent transcription or cause changes in the *rate* at which transcription is initiated (see text p. 597). Therefore, unless an operator site is affected and repressor attachment cannot occur, mutation at a promoter site would not be expected to cause constitutive enzyme synthesis.

29-5. The proposed model can be pictured as follows [each operon contains a regulator gene (RG) as well as a number of structural genes (SG)]:

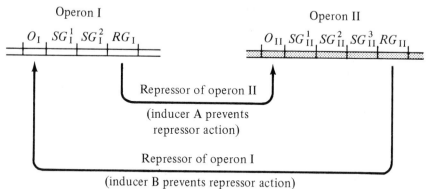

The regulator gene of each operon produces a repressor protein that affects the operator gene (O) of the other operon. Thus, activity of RG_I blocks O_{II} and subsequent synthesis of the proteins produced by $SG_{II}^1 \, SG_{II}^2 \, SG_{II}^3 \, RG_{II}$. Similarly, RG_{II} activity blocks O_I and prevents protein synthesis by $SG_I^1 \, SG_I^2 \, RG_I$. Since the two repressors produced by these regulators are each affected by different inducers (RG_I repressor is made inactive by inducer A, and RG_{II} repressor is inactivated by inducer B), the first inducer to which this system is exposed will determine which set of structural genes will subsequently function. Therefore, if inducer A is introduced first, the repressor of operon II is inactivated, thus allowing operon II structural genes ($SG_{II}^1 - SG_{II}^3$) and regulator gene (RG_{II}) to be expressed. As a consequence of RG_{II} expression, operon I is repressed (including repression of RG_I), and the transcription of operon II continues even after inducer A is removed. On the other hand, if inducer B is introduced first, operon I can now function, leading to continued repression of operon II.

(Reference: J. Monod and F. Jacob, 1961, *Cold Spring Harb. Symp.*, **26**:389-401.)

29-6. As shown in the following diagram, a product of each operon produces or acts as an inducer for the other. Thus both operons remain repressed together, and when one begins to function (e.g., perhaps because of an external inducer), the other operon begins to function immediately after. For example, the introduction of an inducer that blocks the activity of the operon I repressor will allow that operon to produce product I. This operon I product acts as an inducer that derepresses operon II, allowing it to make product II. Product II, in turn, acts as an inducer to derepress operon I, and so on.

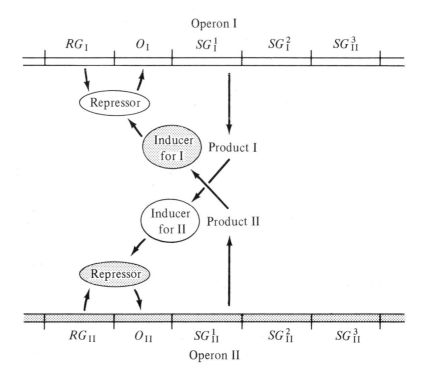

Operon I

Operon II

29-7. Such a mutation would cause virulence of λ by preventing the binding of cI repressor to the operator at the p_L site. This would enable synthesis of the N gene product, followed by activation of late genes by the Q product, and subsequent lysis.

29-8. The *lac* operon is obviously sensitive to catabolite repression since the production of β galactosidase, controlled by the *lac* operon in this strain (*lac* I^+ P^+ O^+ Z^+), is significantly diminished in glucose medium. However, the purine operon does not appear to be at all sensitive to catabolite repression since the production of acetylase, controlled by *pur E* in the F′ episome, is not diminished in the presence of glucose. (Note that, as expected, acetylase production is greatly reduced in the presence of adenine.)

(Reference: B. Magasanik, 1970, *The Lactose Operon*, J.R. Beckwith and D. Zipser, eds. Cold Spring Harbor Laboratory, Cold Spring Harbor, pp. 189-219.)

30

Gene Manipulation

30-1. Since the four bases are equally frequent and randomly distributed, the chances for a particular base, G, C, A, or T, to be found at a nucleotide position is .25. [A particular dinucleotide sequence will thus have a probability of $(.25)^2 = .00625$; meaning that an average nucleotide length (fragment size) of the reciprocal of this number, $1/.00625 = 16$ nucleotides long, can be expected to have one such dinucleotide sequence.]

(a) *Hpa*II recognizes the tetranucleotide sequence CCGG, and such a sequence would be expected to occur in this DNA with a probability of $(.25)^4 = .0039$, or once in an average fragment size of $1/.0039 = 256$ nucleotides long.

(b) *Eco*RI recognizes a six-nucleotide length sequence, GAATTC, which would be expected to occur in this DNA once in an average fragment size of $1/(.25)^6 = 1/.000244 = 4096$ nucleotides long.

30-2. The base ratios given indicate that the probability of a G (or a C) at any nucleotide position is .4, and the probability of an A (or a T) is .1.

(a) Since *Hpa*II recognizes CCGG, such a sequence has a probability of $(.4)^4 = .0256$. That is, a fragment size of $1/.0256 = 39$ nucleotides long, will probably have a CCGG sequence.

(b) The GAATTC sequence recognized by *Eco*RI has a probability of occurring in this DNA $(.4)^2 (.1)^4 = .000016$, or once in a fragment size $1/.000016 = 62,500$ nucleotides long.

30-3. Subject the DNA to degradation by each enzyme and compare the degree of fragmentation produced. If cytosine is not methylated, *Msp*I and *Hpa*II should produce about the same degree of fragmentation. if cytosine is methylated, it is likely that methylation will be present in at least some of these target sequences, and fewer fragments will be produced by *Hpa*II.

30-4. (a) *Xho*I $\left(\begin{array}{l} \cdots \text{C} \downarrow \text{TCGA G} \cdots \\ \cdots \text{G AGCT} \uparrow \text{C} \cdots \end{array}\right)$ and *Sal*I $\left(\begin{array}{l} \cdots \text{G} \downarrow \text{TCGA C} \cdots \\ \cdots \text{C AGCT} \uparrow \text{G} \cdots \end{array}\right)$

(b) *Mbo*I $\left(\begin{array}{l} \cdots \downarrow \text{GATC} \cdots \\ \cdots \text{CTAG} \uparrow \cdots \end{array}\right)$

30-5. (a) Two fragments, 1.8 kb and 1.4 kb long

      (b) Two fragments, .7 kb and 2.5 kb long

      (c) Three fragments, .7 kb, 1.9 kb, and .6 kb long

30-6. Since *Bgl*II digestion produces only two fragments, there is only a single *Bgl*II site (B) on the 3.1-kb sequence, located about one third the distance from one of the ends:

*Eco*RI cuts the sequence twice, producing three fragments that can be mapped in various ways:

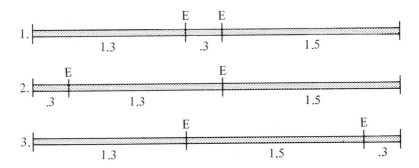

If, let us say, No. 1 above is the proper map for *Eco*RI sites, then the combined *Bgl*II + *Eco*RI digestion should produce four fragments of sizes 1.1 kb, .2 kb, .3 kb, and 1.5 kb:

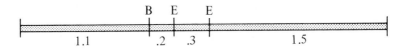

Obviously, this does not fit the data obtained for *Bgl*II + *Eco*RI digestion, nor do any other *Eco*RI maps fit except that of No. 2, indicating that the map must be:

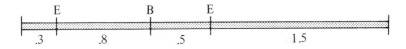

30-7. A simple procedure would be to treat plasmid **1** with *Bgl*II and *Bam*HI, which cut the *drg¹* locus at each end, and splice this fragment into plasmid **2** which has been treated with *Bam*HI. Note that according to text Table 30-1

(see also Problem 30-4), the *Bgl*II-treated end of *drg¹* $\left(\begin{array}{l}\cdots \text{A-3}' \\ \cdots \text{TCTAG-5}'\end{array}\right)$ is complementary to one of the cleavage ends at the *Bam*HI-treated site of plasmid 2 $\left(\begin{array}{l}5'\text{-GATCC}\cdots \\ 3'\text{-G}\cdots\end{array}\right)$. The result will be a plasmid that possesses both drug-resistant genes, one of which (*drg¹*) has a unique *Eco*RI site that can then be used for insertional inactivation.

30-8. One possible procedure is as follows:

1. Eliminate the *Eco*RI site by first digesting the plasmid with *Eco*RI and then use S1 nuclease to remove the exposed *Eco*RI single strands, thereby producing blunt ends (see text Fig. 30-5):

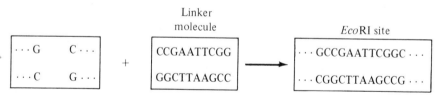

2. Using T4 DNA ligase, the plasmid can be recircularized by blunt-end ligation, but the former *Eco*RI site will now be absent.

3. This new plasmid can be cloned and then subjected to *Bam*HI digestion, followed again by S1 nuclease treatment to produce blunt ends:

4. Synthetic *Eco*RI linker molecules (see text Fig. 30-2) can then be ligated to the blunt ends at the former *Bam*HI site:

The final appearance of the plasmid will therefore be as follows:

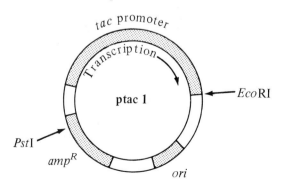

31

Differentiation and Pattern

31-1. One possible explanation is that the mRNA necessary for the production of cap proteins is present in algal cytoplasm at an early stage but is repressed by the presence of a nuclear regulator. Removal of the nucleus removes the regulator and thereby removes the repressor substance, consequently allowing cap formation.

(Reference: G. Werz, 1965, *Brookhaven Symp. Biol.*, **18**:185–202.)

31-2. One possible explanation is that the *o/o* maternal effect is caused by the inability to produce a necessary protein and that the functioning of this protein is dependent on its direct synthesis in the cell. Thus the injected nuclear material that corrects the *o/o* effect may be wild-type nuclear mRNA, which is then capable of directing necessary protein synthesis in the cytoplasm. Another possible explanation is that protein synthesis in abnormal eggs is repressed because of the absence of an inducer from the nucleus, and the injection of wild-type nuclear material provides this inducer.

(Reference: R. Briggs and G. Cassens, 1966, *Proc. Nat. Acad. Sci.*, **55**:1103–1109.)

31-3. Direct maternal transmission of xanthine dehydrogenase enzymes seems excluded as a cause for the appearance of wild-type offspring since homozygous *rosy* females do not produce such enzymes. It seems more likely that the maternal effect is caused in this case by the maternal transmission of a wild-type product from the *maroon-like*$^+$ gene (the mother is *mal*$^+$/*mal*), which then interacts with the wild-type product of the *rosy*$^+$ gene since all of the progeny are *ry*$^+$/*ry*. The molecular nature of this interaction is not yet known.

(See E. Glassman, 1965, *Fed. Proc.*, **24**:1243–1251).

31-4. The experiments demonstrate that the activity of injected brain nuclei parallels the activity of nuclei normally present in the recipient cells. In terms of the three results described:

   (a) Developing oocytes produce considerable amounts of RNA to serve for both ribosomal RNA and messenger RNA purposes in later fertilized eggs.

(b) As the oocyte matures, RNA synthesis ceases, and both the oocyte nucleus and its chromosomes condense as they undergo meiosis.

(c) The fully mature fertilized egg will normally undergo a number of DNA replications before RNA synthesis resumes.

The precise control exercised by the cytoplasm over nuclear activity can therefore extend to injected nuclei as well.

(Reference: J.B. Gurdon, 1968, *J. Embryol. Exp. Morphol.*, **20**:401–414.)

31-5. (a) If LDH was a trimer composed of three subunits, the number of different phenotypes would be four; that is, 3α, $2\alpha{:}1\beta$, $1\alpha{:}2\beta$, and 3β. Note that since there are only two kinds of subunits, α and β, the number of different phenotypes is equal to the number of groups produced by a binomial expansion, $(\alpha + \beta)^n$, where n represents the number of subunits that compose one LDH isozyme. As shown in text Table 8-6, the number of such groups is $n + 1$. Thus, if an enzyme was composed of six subunits, each subunit either α or β, the number of different possible isozymes would be seven. (As also shown in Table 8-6, these different isozymes would be present in the ratio $1{:}6{:}15{:}20{:}15{:}6{:}1$, assuming that equal amounts of the α and β subunits are introduced, and that there is random combination between all subunits to produce the various six-subunit isozymes.)

(b) If there are three possible kinds of subunits, the number of different possible phenotypes equals the number of groups produced by a trinomial expansion, $(\alpha + \beta + \gamma)^n$, where n is again the number of subunits that compose one LDH isozyme. In the case of a trinomial, there are $(n + 1)(n + 2)/2$ different groups produced, and therefore $(5)(6)/2 = 15$ possible isozyme phenotypes, given that three kinds of subunits compose a tetrameric enzyme.

These examples indicate that relatively few genes can produce a large number of different kinds of isozyme within a single individual if the isozyme can be composed of more than one kind of polypeptide chain. As pointed out on text pp. 635–636 and Fig. 31-5, the presence of such different isozymes provides an important source of biochemical variability to individual organisms during their own differentiation.

31-6. If the appearance of the two enzymes were regulated only by transcriptional controls, it would of course be possible that the two genes are in the same operon subject to the same operator. It is also possible, however, that they are in different operons whose operators are controlled by the same inducer, or perhaps controlled by different inducers that appear simultaneously. On the other hand, translational controls can also account for the synchronous appearance of two different enzymes if translation of the two different mRNAs is subject to the same initiation signals. By the same reasoning, some coordinated control over the two enzymes can also be exercised at the posttranslational level by means of protease degradation that acts on these two enzymes simultaneously.

31-7. The third curve in text Figure 31-34 (□—□—□) shows clearly that RNA synthesis is continuous in the duck erythroblasts during the incubation period although the total amount of RNA present at any one time is not more than 600 to 800 counts per minute. Thus it is *not* the cessation of RNA synthesis that accounts for the plateau shown in the first curve (○—○—○). Rather, it seems more likely that most of the RNA synthesized is *unstable* and that new RNA is being synthesized as older RNA is degraded. This view is supported by the fact that actinomycin D treatment (△—△—△) causes a reduction in the amount of RNA synthesized to about 20 percent. This indicates that the prevention of transcription removes about 80 percent of the expected RNA, or that only about 20 percent is stable and still remains from the period prior to actinomycin treatment.

[Various further studies have indicated that the unstable portion is the RNA that remains in the nucleus, whereas the synthesized RNA sent to the cytoplasm is associated with polysomes (mostly hemoglobin mRNA) and is fairly stable. (See G. Attardi, H. Parnas, M.I.H. Hwang, and B. Attardi, 1966, *J. Mol. Biol.*, **20**:145-182.)]

31-8. Since the chromosomes in animal sperm are generally nonfunctional, there is no need for regulation of their activity, and a simple general repressor (i.e., protamine) may be a satisfactory substitute for the more complex repressors used in most cellular regulation.

31-9. According to Lewis's model of bithorax complex activity (see text Fig. 31-32), the presence of *Ubx*$^+$ is responsible for the differentiation of structures in segment T3 (halteres). Presumably, *Ubx*$^+$ activity is normally repressed in segment T2 (wings). One explanation of *Contrabithorax* activity is therefore that *Cbx*$^+$ aids or allows *Ubx*$^+$ repression in segment T2, and the *Cbx* mutation derepresses *Ubx*$^+$ activity in that segment. Thus, T2 structures are transformed into T3 structures in the *Cbx* mutant.

31-10. The fate-map distance derives from the frequency in which a "mosaic boundary" (see text p. 656) passes between two developmental sites. In the present example there are two classes of phenotypes that show such boundaries between the drop-dead effect and the abdominal surface: gynandromorphs that are normal for drop-dead but mutant for the abdomen, and gynandromorphs that show the drop-dead effect but are normal for the abdomen. Since the frequency of these two classes of flies is $(28 + 23)/(54 + 28 + 23 + 31) = 51/136 = .38$, the drop-dead-abdomen fate-map distance is 38 "sturts."

32
Gene Frequencies and Equilibrium

32-1. The percentages given in the problem are phenotypic frequencies in the population and not ratios produced by a particular set of crosses. Thus, it is entirely possible that black stripes are caused by a recessive allele whose frequency (q) is $\sqrt{.75} = .87$, and the absence of such stripes caused by a dominant allele whose frequency (p) is $1 - .87 = .13$. As discussed on text p. 670, a dominant allele need not be present in high frequency.

32-2. $q = \sqrt{.09} = .30$; $p = .70$; $2pq = .42 =$ frequency of Aa heterozygotes.

32-3. $p^2 + 2pq = .19$; $q^2 = .81$; $q = .9$; $p = .1$; therefore $2pq = .18 =$ frequency of Aa heterozygotes.

32-4. (a) A mating between two normal, nonrelated individuals will produce an alkaptonuric offspring only when the two parents are heterozygotes. The frequency of such matings is $2pq \times 2pq = 2(.999)(.001) \times 2(.999)(.001) = .000004$, and the frequency of affected offspring that result from such matings is .25. Thus the overall frequency of such an event would be $.000004 \times .25 = 1 \times 10^{-6}$.

    (b) The alkaptonuric parent is, by definition, a homozygous recessive, and the chances for an alkaptonuric offspring to arise from this mating therefore depends on whether the other parent is a heterozygote. Since the frequency of a heterozygote in this population is $2pq = 2(.999)(.001) = .002$, and the frequency of an affected offspring to arise from such a mating is .50, the total probability of the event is therefore $.002 \times .50 = .001$.

    (c) A normal individual with an affected sibling has a 2/3 probability of being a heterozygote since his normal parents (who are therefore heterozygotes) produce normal offspring in the ratio of 1 homozygous normal to 2 heterozygous normal (see answer to Problem 8-2, part b). Since the chances that a normal, nonrelated individual in the population is a heterozygote is .002, the chances that such a mating will produce an affected offspring is $2/3 \times .002 \times .25 = .00033$.

32-5. (a) .441   (b) .445   (c) .438   (d) .465   (e) .490

(Reference: B. Broman, A. Heiken, and J. Hirschfeld, 1963, *Acta Genet. Stat. Med.*, **13**:132-139.)

32-6. In both populations the frequency of A (p) = .4, and a (q) = .6. The equilibrium genotypic frequencies therefore will be .16 AA, .48 Aa, .36 aa.

32-7. (a) Populations II and III.
　　　(b) I: .450 AA, .442 Aa, .108 aa. IV: .0102 AA, .1816 Aa, .8082 aa.
V: .0025 AA, .0941 Aa, .9034 aa.
　　　(c) One generation.

32-8. Z (p) = $(1084 + 1043)/4094 = .52$; z (q) = $(924 + 1043)/4094 = .48$; p^2 (ZZ) = .27, $2pq$ (Zz) = .50, q^2 (zz) = .23. Multiplying each of these frequencies by the total number (2047) gives expected numbers of 553 ZZ, 1023 Zz, 471 zz. A chi-square test yields a value of .782, which at 1 degree of freedom is certainly nonsignificant, and we can therefore accept the hypothesis of Hardy-Weinberg equilibrium.

(Reference: C. Stormont, 1952, *Genetics*, 37:39-48.)

32-9. (a) Frequency $M = 662/722 = .917$; $N = 60/722 = .083$. Expected values of $MM = (.917)^2 (361) = 304$; $MN = 2(.917)(.083)(361) = 55$; $NN = (.083)^2 (361) = 2$. Chi-square at 1 degree of freedom = .667. The hypothesis of Hardy-Weinberg equilibrium is therefore accepted.

(From data of W.C. Boyd, 1950. *Genetics and the Races of Man.* Little, Brown, Boston, p. 234.)

　　(b) The frequency of the N gene in the male population is .083. Assuming random mating, a woman of N phenotype therefore will have a .083 chance of producing a child with N phenotype.
　　(c) Whatever mating occurs, half the children of a heterozygote will be heterozygous. The answer is therefore .5.

32-10. To answer this problem, a 3 × 2 contingency chi-square can be calculated as follows:

Observed

	MM	*MN*	*NN*	Totals
Navaho	305	52	4	361
Pueblo	83	46	11	140
Totals	388	98	15	501

Expected

	MM	*MN*	*NN*	Totals
Navaho	$(388/501)(361) = 279$	$(98/501)(361) = 71$	$(15/501)(361) = 11$	361
Pueblo	$(388/501)(140) = 109$	$(98/501)(140) = 27$	$(15/501)(140) = 4$	140
Totals	388	98	15	501

Observed − Expected

	MM	*MN*	*NN*
Navaho	26	−19	−7
Pueblo	−26	19	7

$\chi^2 = \Sigma(\text{obs} - \text{exp.})^2/\text{exp.} = 2.423 + 5.084 + 4.454 + 6.202 + 13.370 +$ 12.250 = 43.783, which at 1 degree of freedom shows that there is a significant difference between the populations.

(From data of Boyd, p. 223; see answer to Problem 32-9a.)

32-11. Frequencies: O = .777; A = .223; M = $(.917)^2$ = .841; MN = 2(.917) (.083) = .152; N = $(.083)^2$ = .007.

(From Boyd, p. 234; see answer to Problem 32-9a.)

    (a) NO = (.007)(.777) = .005; MNO = (.152)(.777) = .118; MA = (.841)(.223) = .188.
    (b) Linkage would slow the approach to equilibrium but would not change the equilibrium values.
    (c) As in (b).
    (d) Equilibrium values would be reached with the same speed as unlinked loci.

32-12. Only an MM father would be cleared by such a test, and the probability of such individuals in the population is p^2.

32-13. Determine the equilibrium frequencies of the kinds of gametes that are involved in producing the desired genotype (see text p. 673). In the present case, the *aabb* genotype is determined by the union of *ab* × *ab* gametes. The gametic frequencies, in turn, depend on the gene frequencies of *a* and *b*. The answers are therefore as follows:
    (a) Since the frequencies of both *a* and *b* in this population are .3, the frequency of *ab* gametes at equilibrium will be (.3)(.3) = .09. Thus the *aabb* genotype formed by the union of two *ab* gametes will be in expected frequency of $(.09)^2$ = .0081.
    (b) In this population the frequency of *a* is .7, and *b* is .3. The frequency of *ab* gametes are therefore (.7)(.3) = .21, and the frequency of *aabb* genotypes are $(.21)^2$ = .0441.
    (c) By similar reasoning as above, *a* = .5; *b* = .5; *ab* = .25; *aabb* = .0625.
    (d) *a* = .8; *b* = .2; *ab* = .16; *aabb* = .0256

32-14. As discussed on text pp. 674-675, the rate at which genotypic equilibrium is reached depends on the size of the disequilibrium factor *d* or the difference in proportion between coupling and repulsion gametes [(*AB*)(*ab*) − (*aB*)(*Ab*)]. If this factor is initially large, then genotypic equilibrium will be reached more slowly than if it is small. For the populations given in Problem 32-13, d is as follows:

  (a) = .21
  (b) = .21
  (c) = .25
  (d) = .16

Thus, population (c) is furthest from genotypic equilibrium.

32-15. Number of $tt = 569 - 442 = 127$; frequency $tt = 127/569 = .223 = q^2$; $q = \sqrt{.223} = .473$; $p = .527$.

(From data of Boyd, p. 280; see answer to Problem 32-9a.)

  (a) $T\text{-} \times T\text{-} = (p^2 + 2pq)^2 = (.777)^2 = .604$.
  (b) $q^2/(1 + q)^2 = (.473)^2/(1.473)^2 = .223/2.17 = .103$ nontasters, or a ratio of 8.7 tasters:1 nontaster.
  (c) $q/(1 + q) = .473/1.473 = .321$ nontasters, or a ratio of 2.12 tasters:1 nontaster. (The expected ratios for (b), (c) are derived on text p. 679.)

32-16. (a) The corrected gene frequency estimates obtained according to the method on text p. 682 $(d = .0148)$ are $p\,(A) = .2756$, $q\,(B) = .1907$, $r\,(O) = .5336$. These estimates give the following expected numbers of blood types, $O = (r^2)(502) = 143$; $A = (p^2 + 2pr)(502) = 186$; $B = (q^2 + 2qr)(502) = 120$; $AB = (2pq)(502) = 53$. Chi-square at 1 degree of freedom is calculated as 2.577 and indicates no significant difference from Hardy-Weinberg equilibrium.
  (b) $(.777)(.105) = .0816$
  (c) $(.777)(.2847) = .2212$

32-17. (a) Frequency $a = \dfrac{2A + AB + AC}{2N} = \dfrac{312}{358} = .872 = p$

    Frequency $b = \dfrac{2B + AB + BC}{2N} = \dfrac{13}{358} = .036 = q$

    Frequency $c = \dfrac{2C + AC + BC}{2N} = \dfrac{33}{358} = .092 = r$

(b)

Phenotype	Observed No.	Expected No.
A	134	$p^2 \times 179 = 136.1$
AB	11	$2pq \times 179 = 11.2$
B	1	$q^2 \times 179 = .2$
AC	33	$2pr \times 179 = 28.7$
BC	0	$2qr \times 179 = 1.2$
C	0	$r^2 \times 179 = 1.5$
	$N = 179$	

The numbers expected according to Hardy-Weinberg equilibrium are obviously in accord with those observed.

(Reference: L.Y.C. Lai, 1967, *Acta Genet. Stat. Med.*, **17**:104-111.)

32-18. Frequency $A = \dfrac{2A + AB + AC + AD}{2N} = \dfrac{48}{226} = .212 = p$

Frequency $B = \dfrac{2B + AB + BC + BD}{2N} = \dfrac{175}{226} = .774 = q$

Frequency $C = \dfrac{2C + AC + BC + CD}{2N} = \dfrac{2}{226} = .009 = r$

Frequency $D = \dfrac{2D + AD + BD + CD}{2N} = \dfrac{1}{226} = .005 = s$

Phenotype	Observed No.	Expected No.
A	4	$p^2 \times 113 = 5.1$
AB	38	$2pq \times 113 = 37.1$
B	68	$q^2 \times 113 = 67.7$
AC	1	$2pr \times 113 = .4$
BC	1	$2qr \times 113 = 1.6$
AD	1	$2ps \times 113 = .2$
BD	0	$2qs \times 113 = .008$
C	0	$r^2 \times 113 = .01$
D	0	$s^2 \times 113 = .003$
CD	0	$2rs \times 113 = .01$
	$N = 113$	

As in the previous problem, the numbers expected according to Hardy-Weinberg equilibrium are very much in accord with those observed.

(Reference: E.A. Azen and O. Smithies, 1968, *Science*, **162**:905-907.)

32-19. (a) $p(A) = .2841$, $q(B) = .0270$, $r(O) = .6889$

(b) Expected blood type numbers: $O = (r^2)(600) = .285$; $A = (p^2 + 2pr)(600) = 283$; $B = (q^2 + 2qr)(600) = 23$; $AB = (2pq)(600) = 9$

Chi-square at 1 degree of freedom is 2.537. These data therefore show no significant differences from Hardy-Weinberg equilibrium.

(Reference: L.P. Chiasson, 1963, *J. Hered.*, **54**:229-236.)

32-20. The expected frequency of O phenotypes would be $q^3 = (.7)^3 = .34$. Actual data have shown that individuals with Down syndrome do not have O blood type frequencies different from those who are disomic for chromosome 21. Because of these findings and other data, the ABO blood group locus is not believed to be located on chromosome 21.

32-21. (a) Since the baldness allele (e.g., B^1) is dominant in males, the 51 percent frequency of bald men means that this value includes p^2 (B^1B^1) + $2pq$ (B^1B^2). The frequency of homozygous "recessives," q^2 (B^2B^2), is therefore $1 - .51 = .49$, and the frequency of the nonbaldness allele (B^2) is $\sqrt{.49} = .7$. Since the nonbaldness B^2 allele acts as a dominant in females (frequency .7) and the baldness B^1 allele acts as a recessive (frequency $1 - .7 = .3$), the expected frequency of bald women (B^1B^1) is $(.3)^2 = .09$.

(b) $.51$ (B^1-) ♂ × $.91$ (B^2-) ♀ = $.46$

(c) Since the B^1 baldness allele is dominant in males, we can only be *certain* that a male child will be bald if *one or both* of his parents are homozygous for this allele. As calculated in part (a), the frequency of B^1 is .3 and the frequency of B^1B^1 baldness homozygotes is $(.3)^2 = .09$, or 9 percent of males and 9 percent of females. No matter with whom these groups of 9 percent mate, they are certain to pass on a baldness allele to their offspring. That is, the male children of 9 percent of random matings in this population are certain to develop baldness. On the other hand, we can only be certain that a bald daughter (B^1B^1) will be produced when *both* parents are B^1B^1 homozygotes, that is, $.09 \times .09 = .0081$, or in less than 1 percent of all matings.

(d) Since both parents are nonbald, the father must be B^2B^2, and a male child will only develop baldness if the bald B^1 allele is transmitted to him by a heterozygous (B^1B^2) mother. The frequency of such heterozygotes ($2pq$) among nonbald females $(1 - q^2)$ is $(2pq)/(1 - q^2) = .42/.91 = .46$. Since matings between heterozygous females and nonbald males produce male offspring who have a .5 chance of being bald, the total probability of two nonbald parents producing a bald son is $.46 \times .50 = .23$.

(e) If the mother is bald, the chances for the daughter to be bald will depend on the probability of the bald allele in the gamete contributed by the unknown father. This probability is equal to the frequency of the baldness allele in this population, or .3.

32-22. There are four alleles whose frequencies can be designated as: $MS = p = 54/262 = .2061$, $Ms = q = 99/262 = .3779$, $NS = x = 16/262 = .0611$, $Ns = y = 93/262 = .3550$. (The MS/Ns or Ms/NS phenotypic data cannot be used in gene frequency calculations since the exact genotypes are unknown.) On this basis, the expected frequencies and numbers of genotypes at Hardy-Weinberg equilibrium are $MS/MS = p^2 = 7.39$, $MS/Ms = 2pq = 27.10$, $MS/NS = 2px = 4.38$, $MS/Ns + Ms/NS = 2py + 2qx = 33.50$, $Ms/Ms = q^2 = 24.85$, $Ms/Ns = 2qy = 46.69$, $NS/NS = x^2 = .65$, $NS/Ns = 2xy = 7.55$, $Ns/Ns = y^2 = 21.93$. Chi-square calculated from these expected numbers yields a value of 14.49, which at 5 degrees of freedom (9 classes $-$ 4 alleles) represents a significant departure from the Hardy-Weinberg equilibrium at the .05 level.

(Reference: A. Heiken, 1965, *Hereditas*, 53:187-211.)

32-23. The gene frequencies are $Se = p = 105/324 = .324$ and $se = q = 219/324 = .676$. The expected frequencies of matings (see Tables 32-4 and 32-8 in the text) are for the respective order of matings given in the problem: .88, 7.42, 15.52, 7.76, 32.40, 16.90. Calculation of chi-square yields a value of 5.88, which at 5 degrees of freedom indicates no significant departure of mating frequencies from those expected at Hardy-Weinberg equilibrium.

32-24. The incompatible pairs will involve only homozygous recipients since the heterozygotes can accept tissue from any of the three genotypes. For example, an A^1A^1 recipient will accept tissue from other homozygotes like itself but will reject tissue from A^2A^2 homozygotes and A^1A^2 heterozygotes. Thus, incompatibility for A^1A^1 recipients is present in the proportion $1/4\,(A^2A^2) + 2/4\,(A^1A^2) = 3/4$ of the time if Hardy-Weinberg equilibrium has been attained. Since the A^1A^1 genotype itself is present in the population in $1/4$ frequency, this means that it causes incompatibility in $3/4 \times 1/4 = 3/16$ of all tissue grafts between random individuals. By similar reasoning, the A^2A^2 genotype causes incompatibility in $3/16$ of all random combinations of tissue grafts, and the frequency of incompatible combinations in this population is therefore $3/16 + 3/16 = 3/8 = 37.5$ percent.

32-25. The only incompatible parent-child combinations in this population are those in which the child is a homozygote and the parent is a heterozygote. (If the parent is also a homozygote, then the child must be compatible since it has the parental genotype.) Thus A^1A^2 parental tissue will be rejected by A^1A^1 children. In the given population, heterozygotes are present in $1/2$ frequency ($2pq = 2 \times .5 \times .5$), and produce homozygous A^1A^1 children in $1/4$ frequency from all of their possible matings [$A^1A^2 \times A^1A^2 \rightarrow 1/4\,A^1A^1$; or $A^1A^2 \times (A^1A^1)$ and $(A^2A^2) \rightarrow 1/4\,A^1A^1$]. Thus the presence of A^1A^1 children will lead to parent-child incompatibilities in this population in frequency $1/2 \times 1/4 = 1/8$. Since A^2A^2 children account for similar incompatibility ($1/8$), the total parent-child incompatibility in this population is $1/8 + 1/8 = 1/4$.

32-26. An increased number of alleles for a histocompatibility gene would be expected to increase the variety of allelic combinations and thereby increase the chances that individuals will differ from each other (e.g., $A^1A^2, A^2A^2, A^2A^3, A^3A^4$, etc.). The chances for an individual to be heterozygous is also increased by these means, and incompatible combinations between parents and children will consequently increase because of the greater frequency of heterozygous parents.

32-27. $q = 1/5000 = .0002$. Frequency in females $= q^2 = .00000004$.

32-28. (a) The observed gene frequencies are $A = p = .6$, $a = q = .4$, $B = r = .7$, $b = s = .3$. The expected equilibrium frequencies of male genotypes are $AB = pr = .42$, $AB = ps = .18$, $aB = qr = .28$, $ab = qs = .12$.
   (b) No (see text pp. 675-676). If the two genes are close together on the X chromosome, equilibrium will also be delayed because of linkage.

(c) The double-recessive *ab* frequency in females $= q^2s^2 = (.16)(.09) =$ .0144.

32-29. (a) In females, the gene frequencies are $+ = p = 608/676 = .899$, and $y = q = 68/676 = .101$. In males, these frequencies are $+ = p = 311/353 = .881$, and $y = q = 42/353 = .119$. A chi-square test for independence shows no significant difference between these values ($\chi^2 = .64$). They can therefore be lumped to give a frequency of $y = q = [2(7) + 54 + 42]/(676 + 353) = .107$, and $+$ is therefore .893.

(b) Based on the above frequencies, the expected equilibrium genotype frequencies in females are: $+/+ = p^2 (338) = 269.66$; $+/y = 2pq(338) = 64.46$; $y/y = q^2(338) = 3.85$. Calculation of chi-square, using Yates correction for 1 degree of freedom, gives a value of 3.54, which indicates that the population has not significantly departed from Hardy-Weinberg equilibrium. (If Yates correction term is not used, $\chi^2 = 4.47$, which is significant at the .05 level.)

(Reference: A.G. Searle, 1949, *J. Genet.*, **49**:214-220.)

32-30. The expected frequencies of female phenotypes would be wild type $= (p^2 + 2pq)(338) = 334$, and yellow $= q^2(338) = 4$. Chi-square at 1 degree of freedom with Yates correction equals 1.58—a value that certainly indicates no significant departure from Hardy-Weinberg equilibrium.

32-31. (a) All normal phenotypes in the first generation.

(b) The gene frequency for the colorblind allele (c) is $q = 10/(15 + 30) = .222$, and for the normal allele (C) is $p = .778$. At equilibrium, these will also be the phenotypic frequencies of the males. The female equilibrium frequencies will be $p^2 (CC) = .605$, $2pq (Cc) = .346$, $q^2 (cc) = .049$.

32-32. The observed frequency of colorblindness in males indicates a gene frequency (q) of $725/9049 = .0801$. Assuming the existence of equilibrium, the expected number of colorblind females would be $q^2 \times 9072 = 58$, and the expected number of normal females would be 9014. A chi-square test with 1 degree of freedom yields a value of 5.62, indicating that equilibrium has not been reached.

(Reference: G.H.M. Waaler, 1927, *Zeitschr. Abstram. Vererbung.*, **45**:279-333.)

32-33. Frequency $c^P = \sqrt{3/9072} = .01818$; frequency $c^d = \sqrt{37/9072} = .06386$. This means that the frequency of expected but unobserved $c^P c^d$ heterozygotes equals $2pqN = 2(.01818)(.06386)(9072) = 21$. In other words, the population probably contains 40 observed $+$ 21 unobserved or a total of 61 females carrying two colorblind alleles. Since our previous calculations indicated that we would have expected 58 such females from the frequencies of colorblind genes in males, we can see that the departure from Hardy-Weinberg equilibrium is not as great as it appeared in the last problem. If we consider the observed number of colorblind females as 61 and the observed number of normals as 9011, a χ^2 test (expected numbers: 58 and 9014) yields .11, indicating that we can accept the hypothesis of Hardy-Weinberg equilibrium.

32-34. The observed frequency of nontasters (homozygous recessives, q^2) is $379/1394 = .272$. The nontaster gene frequency, q, is therefore $\sqrt{.272} = .521$. At equilibrium, the expected frequency of nontaster offspring from taster X taster matings would be, according to Snyder's population ratios (text p. 679), $q^2/(1 + q)^2 = .272/1.521^2 = .117$. This compares with an observed frequency for such offspring of $76/730 = .104$. Similarly, the expected frequency of nontaster offspring from taster X nontaster matings would be $q/(1 + q) = .521/1.521 = .342$, and this compares with the observed frequency of $205/559 = .367$. For both ratios, the observed frequencies seem fairly close to those expected, and we can assume that Hardy-Weinberg equilibrium exists.

(Reference: F.B. Hutt, 1964, *Animal Genetics*. Ronald Press, New York, p. 371.)

33
Changes in Gene Frequencies

33-1. $[10^{-6}$ mutation rate$] \times [2 \times 10^7]$ genes $= 20$ mutant genes. The survival of a mutant gene in the first generation is $1 - .3679 = .6321$ (see Table 33-1 in the text). The number of mutant genes surviving is therefore $.6321 \times 20 \cong 13$.

33-2. The probability for extinction is increased since the number of genes of all types is decreased.

33-3. (a) $A = p_0 (1 - u)^n = [1 - (1 \times 10^{-6})]^{10} = .999999^{10} = .9998$
    (b) $A = [1 - (1 \times 10^{-6})]^{100} = .9977$
    (c) $A = [1 - (1 \times 10^{-6})]^{1000} = .9773$
    (d) $A = [1 - (1 \times 10^{-6})]^{10,000} = .7943$
    (e) $A = [1 - (1 \times 10^{-6})]^{1,000,000} = .0000$

33-4. $(u + v)n = 2.303 \log_{10} (q_0 - \hat{q})/(q_n - \hat{q})$
    $(1 \times 10^{-6})n = 2.303 \log_{10} (.1 - 1.0)/(.5 - 1.0) = 2.303 \log_{10} 9/5$
    $(1 \times 10^{-6})n = 2.303 (.25527) = .57889$
        $n = .57889/1 \times 10^{-6} = 578,890$

33-5. $u = 1.3 \times 10^{-4}, v = 4.2 \times 10^{-5}$
    (a) According to text p. 688, $\hat{q} = u/(u + v) = .00013/.000172 = 756$.
    (b) $(u + v)n = 2.303 \log_{10} (q_0 - \hat{q})/(q_n - \hat{q})$
        $(.000172)n = 2.303 \log_{10} (.1 - .756)/(.5 - .756) = 2.303 \log_{10} 2.5625$
        $= 2.303(.40866) = .9411$
        $n = .9411/.000172 = 5472$

33-6. $\hat{p} = v/(u + v) = 1 \times 10^{-6}/(1 \times 10^{-6} + 1 \times 10^{-6}) = 1/2 = .5$

33-7.

	(a)	(b)	(c)
	Haploid	Diploid Dominance	Diploid No Dominance
q	$\dfrac{-sq(1-q)}{1-sq}$	$\dfrac{-sq^2(1-q)}{1-sq^2}$	$\dfrac{-sq(1-q)}{1-2sq}$
.01	.00298	.0000297	.00299
.1	.0278	.00271	.0287
.3	.0692	.0194	.0768
.5	.0882	.0405	.1071
.7	.0797	.0516	.1086
.9	.0370	.0321	.05869
.99	.00422	.00416	.00731

(over Δq heading)

(d)

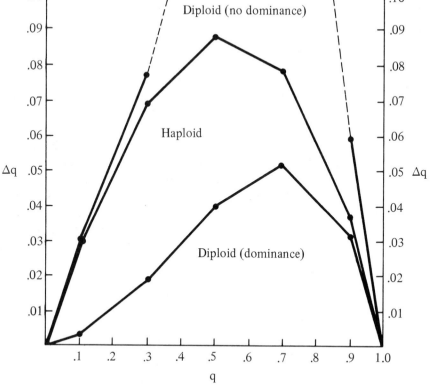

33-8. p (frequency AR) = .2; q (frequency CH) = .8; s = .25; t = .58.

(a) According to text Table 33-12, frequency of q (CH) = q $(1-qt)/(1-p^2s-q^2t)$ = .43/.62 = .69. Therefore, the frequency of p (AR) = .31.

(b) $\hat{q}$ = s/(s + t) = .25/.83 = .30; $\hat{p}$ = t/(s + t) = .58/.83 = .70

(c) The eggs are laid before selection occurs, and we would therefore expect to find Hardy-Weinberg equilibrium among the egg genotypes. For the adult genotypes we would expect the following:

	AR/AR	AR/CH	CH/CH	Total
Preselection frequencies	p^2	2pq	q^2	
(p = .7, q = .3)	.49	.42	.09	1.00
Adaptive value	.75	1.00	.42	
Postselection	.3675	.4200	.0378	.8253
Relative frequencies				
postselection	.445	.509	.046	1.000

Note that the equilibrium gene frequencies are the same for both pre- and post-selection, only the genotypic frequencies change.

(Reference: Th. Dobzhansky and O. Pavlovsky, 1959, *Proc. Nat. Acad. Sci.,* **46**:41–47.)

33-9. p = .80; q = .20; s = .50; t = 1.00

(a) q = q$(1-qt)/(1-p^2s-q^2t)$ = .16/.64 = .25

(b) This .33 frequency is the equilibrium frequency, $\hat{q}$ = s/(s + t) = .50/1.50 = .33, and will maintain itself as long as the selection coefficients do not change.

33-10. (a) We can make the assumption that the adaptive value of the heterozygotes (2pq) is 1.00 and that the departure of the homozygote frequencies from that expected according to the Hardy-Weinberg equilibrium (p^2, q^2) indicates their respective adaptive values. Since the frequency of each gene in the F_1 generation is equal (p = q = .5), we would expect the frequencies of each of the homozygotes to be equal to one half the frequency of heterozygotes (p^2 = q^2 = .25 = 2pq/2 = 3767/2 = 1883.5). The adaptive values of the homozygotes are therefore:

$$EE = 1605/1883.5 = .85, \text{ or } s = .15$$
$$ee = 1310/1883.5 = .70, \text{ or } t = .30$$

(b) $\hat{e}$ = s/(s + t) = .15/.45 = .33

(Reference: A.B. Da Cunha, 1949, *Evolution,* **3**:239–251.)

33-11. (a) Note that there is heterozygote superiority in each of these populations, and equilibrium gene frequencies are therefore $A = \hat{p} = t/(s + t)$, and $a = \hat{q} = s/(s + t)$ (text p. 696). The calculated results are as follows:

Population	s	t	$\hat{p}$	$\hat{q}$
I	.50	1.00	.667	.333
II	.25	.50	.667	.333
III	.10	.20	.667	.333

This indicates quite pointedly that differences in selection coefficients between populations need not lead to differences in equilibrium frequencies as long as s and t maintain proportional relationships to each other.

(b) Although equilibrium frequencies are the same in the three populations, it seems obvious that the larger selection coefficients in population I will cause more rapid gene frequency changes each generation in that population than in the others. If we use the formula

$$\Delta q = \frac{pq\,(ps - qt)}{1 - p^2\,s - q^2\,t}$$

(text p. 696), then the frequency changes in q for the first generation of selection would be as follows:

Population	Original $\hat{q}$ (freq. a)	Δq	q after Selection
I	.600	−.171	.429
II	.600	−.061	.539
III	.600	−.021	.579

33-12. (a)

Gene Frequencies		Genotype Frequencies			
r	R	RR	Rr	rr	Matings $rr \times R-$
.1	.9	.81	.18	.01	$.01 \times .99 = .0099$
.5	.5	.25	.50	.25	$.25 \times .75 = .1875$
.9	.1	.01	.18	.81	$.81 \times .19 = .1539$

(b) Note that such matings will eliminate only heterozygous offspring. The reduction in heterozygotes calculated algebraically, using Table 32-4 in the text (e.g., $T = R$, $t = r$), yields a loss of $p^2\,q^2$ (from matings between $rr\,♀ + RR\,♂$) + pq^3 (from matings between $rr\,♀ \times Rr\,♂$). The frequency of q (r) is therefore reduced to $[2q^2 + pq(2p + q)]/2\,[p^2 + pq(2p + q) + q^2]$. For $r = .1$ the new frequency is therefore $.191/1.982 = .0964$. For $r = .5$, the new frequency is $.875/1.750 = .50$. For $r = .9$, the new frequency is $1.719/1.838 = .9353$.

(c) None of these frequencies will show a stable equilibrium. Note that although there has been no change in frequency when r equals .5, a chance reduction in r will cause a further reduction in r, and a chance increase in r will cause a further increase (see text p. 698).

33-13. (a) The frequency of rh$^-$ mothers (rr) is $q^2 = (.20)^2 = .04$, and the frequency of the R gene among fathers is .8. Thus the frequency of Rh$^+$ children born to such mothers is $.04 \times .8 = .032$.

(b) There are seven mating combinations that are incompatible for ABO blood type: O females $\times$ A, B, or AB males; A females $\times$ B or AB males; B females $\times$ A or AB males. The frequencies of these matings can be calculated from the genotypic frequencies ($O = (.6)^2 = .36$; $A = [(.3)^2 + 2(.3)(.6)] = .45$; $B = [(.10)^2 + 2(.1)(.6)] = .13$; $AB = 2(.3)(.1) = .06$) and yield a total frequency of .3822.

33-14. (a) The equilibrium frequency of the dominant lethal (retinoblastoma) is $p = u/s = 3 \times 10^{-6}$. For the recessive lethal (amaurotic idiocy), $q = \sqrt{u/s} = \sqrt{3 \times 10^{-6}} = 1.73 \times 10^{-3}$.

(b) The expected frequency of homozygotes is $p^2 = 9 \times 10^{-12}$ for the dominant lethal, and $q^2 = 3 \times 10^{-6}$ for the recessive lethal.

33-15. For the dominant gene, equilibrium frequency is $p = u/s = 5 \times 10^{-5}/.667 = 7.50 \times 10^{-5}$. For the recessives, $q = \sqrt{u/s} = \sqrt{3 \times 10^{-5}/.667} = \sqrt{4.5 \times 10^{-5}} = 6.71 \times 10^{-2}$. Note that the equilibrium frequency of a recessive with the same deleterious effect as a dominant ($s = .667$ in both cases) will be much higher even when the mutation rate of the recessive is almost half that of the dominant.

(Reference: C.S. Chung, O.W. Robison, and N.E. Morton, 1959, *Ann. Hum. Genet.*, **23**:357-366.)

33-16. $s = u/p = 1 \times 10^{-6}/1 \times 10^{-3} = 1 \times 10^{-3}$. Adaptive value $= 1 - s = .999$.

33-17. (a) $s = u/p = 1 \times 10^{-6}/1 \times 10^{-6} = 1$

(b) $p = u/s = 2 \times 10^{-6}/1 = 2 \times 10^{-6}$

(c) $q = \sqrt{u/s} = \sqrt{1 \times 10^{-6}/1} = 1 \times 10^{-3}$. A doubling of mutation rate produces $q = \sqrt{2 \times 10^{-6}/1} = 1.414 \times 10^{-3}$.

33-18. If selection against these two genes is reduced to zero ("completely cured"), their frequencies will remain at whatever level they were when selection ceased unless new factors change their frequencies, for example, mutation. If, however, the selection coefficient against these genes was only partially reduced ("partially cured"), there would be a significant difference in the rate at which these two genes were changing. The reason for this stems from the fact that selection is more "efficient" against dominants than against recessives, as discussed on text pp. 692-694. For example, if both recessive and dominant genes begin with the same gene frequency, the recessive phenotype is relatively rarer since it appears only in homozygotes, whereas the dominant phenotype is more frequent because it appears in both homozygotes and heterozygotes. Thus, as shown in the following table, a selection coefficient of $s = .5$ against both dominant and recessive genes that are present in equal frequency $(.01)$ causes a greater change (Δq) in the dominant gene frequency than in the recessive:

	Δq (for $q = .01$)		Reduction in Δq because of Decrease in s (Frequency of "Cured" Genes)
	$s = .5$	$s = .2$	
Dominant $[-sq(1-q)^2]$	.005	.002	.003
Recessive $[-sq^2(1-q)]$	.00005	.00002	.00003

However, note further that when s is reduced to .2 for both these genes, there is a greater reduction in the rate at which the dominant gene frequency is decreasing in relation to the recessive (last column). That is, 3 out of 1000 dominant genes are now "cured" (spared from the effects of the former selection coefficient), whereas only 3 out of 100,000 formerly deleterious recessive genes are cured.

Two conclusions can therefore be drawn: (1) Curing the effects of a deleterious gene would not, by itself, ordinarily increase its frequency in relation to a "normal" gene. (2) Curing the effects of a deleterious dominant gene would have greater effect on a population in terms of numbers of individuals that benefited than curing the effects of a deleterious recessive gene of equal frequency.

33-19. A chi-square test for independence (text Chapter 8) yields a value of 6.15, which is significant at the .05 level, and indicates that polio and PTC-tasting ability are not independent of each other in this sample. From the data it seems clear that there is a greater percentage of tasters (83 percent) among nonpolio patients than among polio patients (69 percent).

(Reference: N. Brand, 1964, *Ann. Hum. Genet.*, 27:233-239.)

33-20. (a) $(1 - m)^n = .89$, or $1 - .89 = .11 =$ proportion of white genes in the Claxton, Georgia, black population, using the Fy^a gene frequency.

(b) $(1 - m)^n = .78$, or $1 - .78 = .22 =$ proportion of white genes in the Oakland, California, black population.

(See T.E. Reed, 1969, *Science*, 165:762-768.)

33-21. $q_0 = .25; Q = .85; m = .05; q_n = (1 - m)^n(q_0 - Q) + Q$
(a) $q_1 = (.95)(-.60) + .85 = .28$
(b) $q_5 = (.95)^5(-.60) + .85 = -.46 + .85 = .39$
(c) $q_{10} = (.95)^{10}(-.60) + .85 = -.36 + .85 = .49$

33-22. $q_0 = 0; q_n = .03; Q = .10; n = 3$

$(1 - m)^n = (q_n - Q)/(q_0 - Q)$

$(1 - m)^3 = .07/.10 = .7$

$1 - m = \sqrt[3]{.7} = .888$

$m = 1 - .888 = .112$

33-23. One possible explanation is that the small amount of migration given in the problem introduced the A blood group gene into the population, and random drift then caused its increase. A factor that would help determine the possibility of random drift would be knowledge of the effective population size.

33-24. (a) $N_e = 4N_f N_m/N_f + N_m) = 4(40)(10)/50 = 32.$ $\sigma = \sqrt{pq/2N} = \sqrt{(.5)(.5)/64} = \sqrt{.0039} = .062.$ The mean and standard deviation is therefore $.500 \pm .062.$

(b) $N_e = 4(10)(10)/20 = 400/20 = 20.$ $\sigma = \sqrt{25/40} = .079.$ The mean and standard deviation is now $.500 \pm .079.$

Note that information on initial gene frequency and effective population size may enable distinctions to be made between genetic drift and other causes for gene frequency change. For example, a population of known size with a particular gene frequency might produce a change in gene frequency because of a chance fluctuation (drift), but this would ordinarily not be expected to be greater than 3 standard deviations in any one generation. (As shown in text Fig. 8-7, the mean $\pm 3\sigma$ includes more than 99 percent of the distribution.) Thus almost any population with the effective size given in part (a) of this problem ($N_e = 32$), and with an initial gene frequency of .50, would rarely be expected to have gene frequencies in the next generation greater than $.50 + 3(.06) = .68$, or less than $.50 - 3(.06) = .32.$ Were a population of this kind found to possess a gene frequency of .80 (a 5σ change!), we could assume that the causes for this rapid frequency charge must include factors other than simple random drift.

33-25. (a) $1/2N = 1/2(32) = 1/64$

$1/64 \times 1000$ populations $= 15.6$ populations per generation

(b) $1/40 \times 1000 = 25$ populations per generation

33-26. Population size data would be necessary. Also, random drift is non-directional (gene frequency changes would be fluctuating), whereas selection is often directional (changes usually tend toward gene elimination or fixation).

33-27. Note that gene frequency changes in the Dunkers have gone in a direction opposite to frequencies in Rhineland Germans and Americans. One possible explanation for this is random drift.

(Reference: B. Glass, M.S. Sacks, E.F. Jahn, and C. Hess, 1952, *Am. Nat.*, 86:145-160.)

34

Inbreeding and Heterosis

34-1. (a) $F = 1/2N = 1/2(75) = .00667$
(b) $F_5 = 1 - (1 - F)^5$ where $F = .00667$ (See footnote text p. 712.)
   $= 1 - (.99333)^5 = 1 - .96711 = .03289$
(c) $F = 1/(2N + 1) = 1/151 = .00662$
(d) $F_5 = 1 - (.99338)^5 = 1 - .96732 = .03268$

34-2. (a) Smaller F than when $N = 75$, since $1/300 < 1/150$.
(b) Larger F than when $N = 75$, since $1/50 > 1/150$.

34-3. $N_e = 4N_f N_m/(N_f + N_m) = 4(40)(10)/(40 + 10) = 1600/50 = 32$
$F = 1/(2N + 1) = 1/65 = .01538$
$F_5 = 1 - (1 - .01538)^5 = 1 - (.98462)^5 = .07457$

34-4. (a) $P_n = P_0(1 - 1/2N)^n$
   $P_{1000} = 1(1 - 1/2000)^{1000} = (.9995)^{1000} = .603$
(b) $F_n = 1 - P_n = 1 - .603 = .397$

34-5. $AA = p^2 + pqF = (.6)^2 + (.6)(.4)(.397) = .36 + .0953 = .4553$
$Aa = 2pq - 2pqF = .48 - (.48)(.397) = .48 - .1906 = .2894$
$aa = q^2 + pqF = .16 + .0953 = .2553$

34-6. (a) $AA = p^2 + pqF = .49 + (.21)(.2) = .49 + .042 = .532$
   $Aa = 2pq - 2pqF = .42 - (.42)(.20) = .42 - .084 = .336$
   $aa = q^2 + pqF = .09 + .042 = .132$
(b) $AA = .49 + (.21)(.5) = .49 + .105 = .595$
   $Aa = .42 - (.42)(.5) = .42 - .21 = .210$
   $aa = .09 + .105 = .195$
(c) $AA = .49 + .21 = .70$
   $Aa = .42 - .42 = 0$
   $aa = .09 + .21 = .30$

34-7. $F = (1/2)^2 \, 1/2 = (1/2)^3 = 1/8 = .125$

34-8. $F = (1/2)^2(1/2)(1 + .4) = (1/2)^3(1.4) = .175$

34-9. Common ancestor B: $F = (1/2)^3 \, 1/2 = (1/2)^4 = .0625$
Common ancestor C: $F = (1/2)^3 \, 1/2 = (1/2)^4 = \underline{.0625}$

$$F = .1250$$

(A is not a common ancestor to H. Note that although G is inbred, it also is not a common ancestor to H.)

34-10. Since neither A nor F is a common ancestor to H, the inbreeding coefficient of H remains the same as calculated for Problem 34-9 (.125).

34-11. We can number the pathways from all common ancestors as follows:

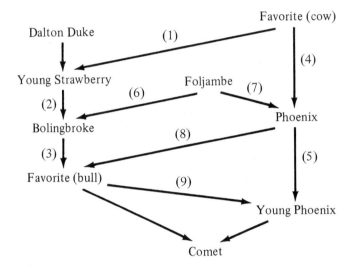

Note that Comet has four common ancestors; Favorite (cow), Foljambe, Phoenix, and Favorite (bull), of which only the latter is inbred. Favorite (bull) has two common ancestors as follows:

Favorite (cow): Path 2 - 1 - 4 $= (1/2)^4 = .0625$
Foljambe: Path 6 - 7 $= (1/2)^3 = \underline{.1250}$

$$F = .1875$$

The inbreeding paths for Comet from each common ancestor are therefore:

Favorite (bull): Path 9 $= (1/2)(1/2)(1 + .1875) = .2969$
Favorite (cow): Path 3 - 2 - 1 - 4 - 5 $= (1/2)^6$ $= .0156$
Foljambe: Path 3 - 6 - 7 - 5 $= (1/2)^5$ $= .0312$
Phoenix: Path 8 - 5 $= (1/2)^3$ $= \underline{.1250}$

$$F = .4687$$

34-12. Sex-linked genes are transmitted on the X chromosome. Since Comet is a male (single X), he cannot be inbred for sex-linked genes.

34-13. Frequency of homozygotes $= 1 - 1/2n = .97; 2n = 1/(1 - .97) = 1/.03 = 33.333 \cong 2^5$. The number of generations is therefore approximately 5.

34-14. The initial frequency of heterozygotes is 1/4 and this is diminished by 1/2 for each generation of self-fertilization. Thus, after five generations of self-fertilization the proportion of heterozygotes remaining is $1/4 \times 1/2^5 = .0078 = .78$ percent.

34-15. Inbreeding changes the genotypic frequencies by increasing the frequency of each of the homozygotes by pqF and reducing the frequency of the heterozygotes by 2pqF. In this problem pq is constant $[(.8)(.2) = .16]$ and the genotypic frequencies are

Condition	F	pqF	AA	Aa	aa
before inbreeding	0	0	.64	.32	.04
(a)	.5	.08	.72	.16	.12
(b)	$1-1/2^5 = .969$	.155	.795	.010	.195
(c)	.25	.04	.68	.24	.08
(d)	.0625	.01	.65	.30	.05

34-16. The genotypic frequencies in males ($A = .8, a = .2$) would not change because of inbreeding. The genotypic frequencies in females, however, would change according to the formulas given on text p. 717, that is, $AA = .64 + (.8)(.2)(.25) = .68, Aa = .32 - 2(.8)(.2)(.25) = .24, aa = .04 + (.8)(.2)(.25) = .08$.

34-17. In the absence of inbreeding the expected frequencies of genotypes are based only on the gene frequencies, which, in this example, are $A = p = .2, a = q = .8$. That is, if F = 0, the expected frequency of $AA = p^2 = .04$. Since the observed frequency of AA is .08 and is caused solely by inbreeding, then $.08 = .04 + pqF$, or $pqF = .08 - .04 = .04$, or $F = .04/pq = .04/(.2)(.8) = .25$.

34-18. (a) $AA = p^2 + pqF = .360 + .096 = .456; Aa = 2pq - 2pqF = .480 - .192 = .288; aa = q^2 + pqF = .160 + .096 = .256$
　　　(b) $AA = p^2 + pqF; .408 = .360 + .240F; .240F = .048; F = .408/.240 = .2$

34-19. In a random mating population $q^2 = .000001$, and the alkaptonuric gene frequency, q, equals $\sqrt{.000001} = .001$. Since inbreeding increases the frequency of alkaptonuric homozygotes by pqF, $.0005 = .000001 + (.999)(.001)F$, or $F = .000499/.000999 = .5$.

34-20. (a) The mating from which individuals 1 and 2 derive is $Aa \times Aa$, and the chance that the wild-type-appearing 2 is a heterozygote is therefore 2/3. If 2 is Aa, there is a 1/2 probability of passing on a, or a probability of $2/3 \times 1/2 = 1/3$ of producing an albino offspring when mating with 1.

(b) There is a 1/2 chance that the maternal parent of 3 is a heterozygote and consequently a 1/2 chance that it will pass on a if that is its genotype. The probability for 3 to be a heterozygote is therefore $1/2 \times 1/2 = 1/4$. If 3 is a heterozygote, the chances for it to pass on a is again 1/2, yielding a total probability of $1/4 \times 1/2 = 1/8$ that a mating between 3 and 1 will produce an albino offspring. (Will this probability change if the first offspring is albino?)

34-21. (a) The aa homozygotes are initially present in frequency $q^2 = (.4)^2 = .16$ and are increased by inbreeding ($F = .0625$) to a frequency of $q^2 + pqF = .16 + (.6)(.4)(.0625) = .175$. The bb homozygotes begin with a frequency of $(.001)^2 = .0000010$ and are increased by inbreeding to $.0000010 + .0000624 = .0000634$.

(b) For aa the relative increase in homozygote frequency is $.175/.16 = 1.09$ [or also $(q + pF)/q = .437/.400 = 1.09$]. For bb the increase is $.0000634/.0000010 = 63.4$

34-22. (a) Using the Snyder ratio for the frequency of recessives produced from dominant $\times$ dominant matings (see text pp. 678-679), $q^2/(1 + q)^2 = (.01)^2/(1.01)^2 = .0001/1.0201 = .000098$.

(b) The relative increase in the frequency of homozygotes in first-cousin matings compared to unrelated matings will be in the ratio of $[q + pF]/q$ (see text p. 717), or $[.01 + (.99)(.0625)]/.01 = 7.19$. (This increase in the probability of recessive homozygotes will actually be less since matings that include homozygous recessive parents have been excluded in this case. Nevertheless the frequencies of recessive homozygotes compared to heterozygotes are so low in this population that their contribution to the production of new recessive homozygotes will detract very little from the given answer.)

34-23. (a) $v(p - q) + 2pqv = .5(.7 - .3) + 2(.7)(.3)(.5) = .41$
(b) $.41 - 2pqvF = .41 - 2(.7)(.3)(.5)(.25) = .3575$

34-24. (a) If the population were completely outbred, F would disappear and the quantitative value (now determined by Hardy-Weinberg equilibrium) would increase to its average value before inbreeding.

(b) Without knowledge of the outbred average value, this observation could be interpreted as overdominance. Actually, only dominance is involved.

34-25. This observation can be interpreted as heterosis caused by overdominance since the yield is greater than that of a normal outbred population with presumed Hardy-Weinberg equilibrium. On the other hand, one could also maintain that the outbred population has nonetheless some inbreeding history and that the inbred self-fertilized lines were homozygous for different deleterious genes, that is, that the "heterotic" hybrids are really carrying dominant genes for more loci than are carried by individuals of the presumed outbred population.

34-26. (a) $(.01)^3$ in egg $\times (.01)^3$ in pollen $= 1 \times 10^{-12}$

(b) At least three generations are required because three crossovers are necessary.

35

Genetic Structure
of Populations

35-1. $S_O = 616/622 = .9904$, $S_S = (1714 - 616)/(1814 - 622) = 9211$; $f_s = (1814 - 622)/1814 = .6571$

$$I = (S_O - S_S) \times f_s = (.9904 - .9211) \times .6571 = .0455$$

(Reference: K. Jayant, 1964, *Ann. Hum. Genet.*, **27**:261-270.)

35-2. $I = (S_O - S_S) \times f_s$;
             $.10 = (S_O - S_S) \times .75$;
   $(S_O - S_S) = .10/.75 = .133$

35-3. The selection intensity is increased since $(S_O - S_S)$ gets larger.

35-4. (a) Yes. Sufficient genetic variability exists in most populations to enable changes in the frequencies of many genes without the need for new mutation. Recombination between existing genes also provides a large source of variation, although limits will be reached eventually.

(b) No. Not all reductions in the variability of a population derive from selection. Inbreeding or random drift can produce homozygosity in a population without a presumed increase in fitness.

35-5. (a) $(.183)(.183) = .0335 = 3.35$ percent
(b) $(1 - .183)^2 (1 - .183)^2 (1 - .143)^2 = 32.72$ percent

35-6. A change in the zone of canalization would be expected to be accompanied by less phenotypic variability than a change confined to choosing only extreme genotypes (see text p. 732).

35-7. It has been suggested that a recombinational event occurred permitting the crossing over and separation of bristle genes that were formerly linked to fertility genes. That is, the "plateau" of bristle numbers reached by the first selection experiment was caused by the fact that further selection for high bristle number was accompanied by the loss of essential fertility alleles that were closely linked to alleles for low bristle number. One or more recombinational events then occurred during relaxed selection, enabling a new linkage between these fertility alleles and alleles for high bristle number.

(Reference: K. Mather and B.J. Harrison, 1949, *Heredity*, 3:1-52, 131-162.)

35-8. (a) $sq^2 = 1(.001) = .001$
(b) $200,000,000 \times sq^2 = 200,000$

35-9. (a) $200,000,000 \times .5(.001) = 100,000$
(b) It will be equal to the mutation rate (see text p. 734).
(c) None (since $sq^2 = u$). (Note, however, that s and q will be different in each case.)

35-10. $2 \times 10^{-7} \times 10r \times 10,000 = 2$ percent deaths at equilibrium

35-11. (a) Load (text p. 735) $= st/(s + t) = (.10)(.59)/.69 = .0855$
(b) $(.0855)(100,000) = 8550$
(This population is also described on text p. 697, and the equilibrium reached by its chromosome arrangements is pictured in Fig. 33-2.)

(Reference: Th. Dobzhansky, and O. Pavlovsky, 1953, *Evolution*, 7:198-210.)

35-12. (a) An increase in the expressed genetic load would be expected because of an increase in the frequency of homozygotes.
(b) As explained on text p. 735, complete inbreeding when homozygotes are superior causes an inbred:outbred load ratio of $1/q$ compared to a ratio of 2 when heterozygotes are superior. Since q is usually small, inbreeding would cause a greater load difference in cases of homozygote superiority than in cases of heterozygote superiority.

35-13. (a) 100 A genes produce 40 A genes in the next generation, and 400 wild-type genes in the normal sibs produce 400 wild-type genes in the next generation (each sib transmits one wild-type gene to each of its offspring). Thus, the relative reproductive success of the A gene compared to the normal gene is $(40/100)/(400/400) = .4$ (see also text p. 699). The selection coefficient against the A gene is therefore $1 - .4 = .6$. Since the mutation rate of A is .00001, the equilibrium frequency is $\hat{p} = u/s = .00001/.6 = .0000167$.
(b) The genetic load caused by a dominant gene is equal to the products of the selection coefficient acting against both dominant homozygotes (p^2) and heterozygotes ($2pq$) or $p^2s + 2pqs$. In the present case p^2 is small enough that it can be ignored without serious consequence, and the load is simply $2pqs = 2(.0000167)(.9999833)(.6) = .0000200$.
(c) $(.00002)(200,000,000) = 4000$

35-14. (a) If the phenylketonuria gene is completely recessive, it is only removed from the population in homozygous condition. Thus a genetic death from this cause is accompanied by the loss of two genes. Since the number of phenylketonuria genes in the population is qN, where $q = \sqrt{q^2} = \sqrt{.001} = .0316$, the total number of genetic deaths that will be caused by phenylketonuria is therefore $qN/2 = .0316N/2$.

(b) If phenylketonuria is treated, these genes will remain in the population for a longer period (more generations) than in the untreated situation, but the total number of genetic deaths will remain the same as long as all phenylketonuria genes must be removed.

35-15. According to the information given, the initial frequency of the advantageous dominant allele is equal to the frequency of the phenylketonuria allele in Problem 35-8; that is,

$$p_0 = q \text{ (phenylketonuria)} = \sqrt{q^2} = \sqrt{.001} = .0316.$$

Calculation of genetic deaths, D (text p. 736), is according to the formula $-2.303 (\log_{10}p_0) = -2.303(-1.5) = 3.454$. [Using natural logarithms: $D = -\ln p_0 = -(-3.454) = 3.454$.]

The total number of genetic deaths expected from this source is therefore about 3-1/2 times the number of individuals normally present in a generation. Note that this loss is less than that for an advantageous dominant allele with lower initial frequency, as given on text p. 736.

35-16. (a) The load produced by the *AA* and *aa* homozygotes is, respectively, p^2s and q^2t (text p. 735). At equilibrium, $p^2s = [t/(s + t)]^2s = (1/1.2)^2\ .20 = .139$, and $q^2t = [s/(s + t)]^2t = (.2/1.2)^2\ 1.00 = .028$. This means that the load caused by *aa* is only $.028/.139$ or about one fifth the load caused by *AA*.

(b) The *AA* frequency $= p^2 = [t/(s + t)]^2 = .694$, and the *aa* frequency $= q^2 = [s/(s + t)]^2 = .028$. As shown above, although *aa* is lethal, the high frequency of *AA* produces a greater proportion of genetic deaths (.139) than does *aa* (.028). Obviously, the relative frequency of a genotype is more important in this instance than its selection coefficient.

35-17. $10^{1000}/10^9 = 10^{991}$

35-18. (a) As the previous answer indicates, there are not enough individuals, let alone populations, to occupy all adaptive peaks.

(b) One possible answer is that some populations are in a process of transition and are presently in a "valley" because of an evolutionary change but will eventually reach a "peak." Another possible explanation is that selection coefficients change as gene frequencies change, so that no matter what adaptive peak a population evolves toward, it continually encounters changed selection coefficients that lead it into "valleys" toward other adaptive peaks. This point was expressed by R.C. Lewontin and M.J.D. White in an examination of grasshopper populations (1960, *Evolution,* **14**:116-129), although other interpretations of their findings are possible.

35-19. (a) Different degrees of genetic drift.

(b) II, since it shows a wide range of gene frequencies.